The Cambridge Manuals of Science and Literature

THE ORIGIN OF EARTHQUAKES

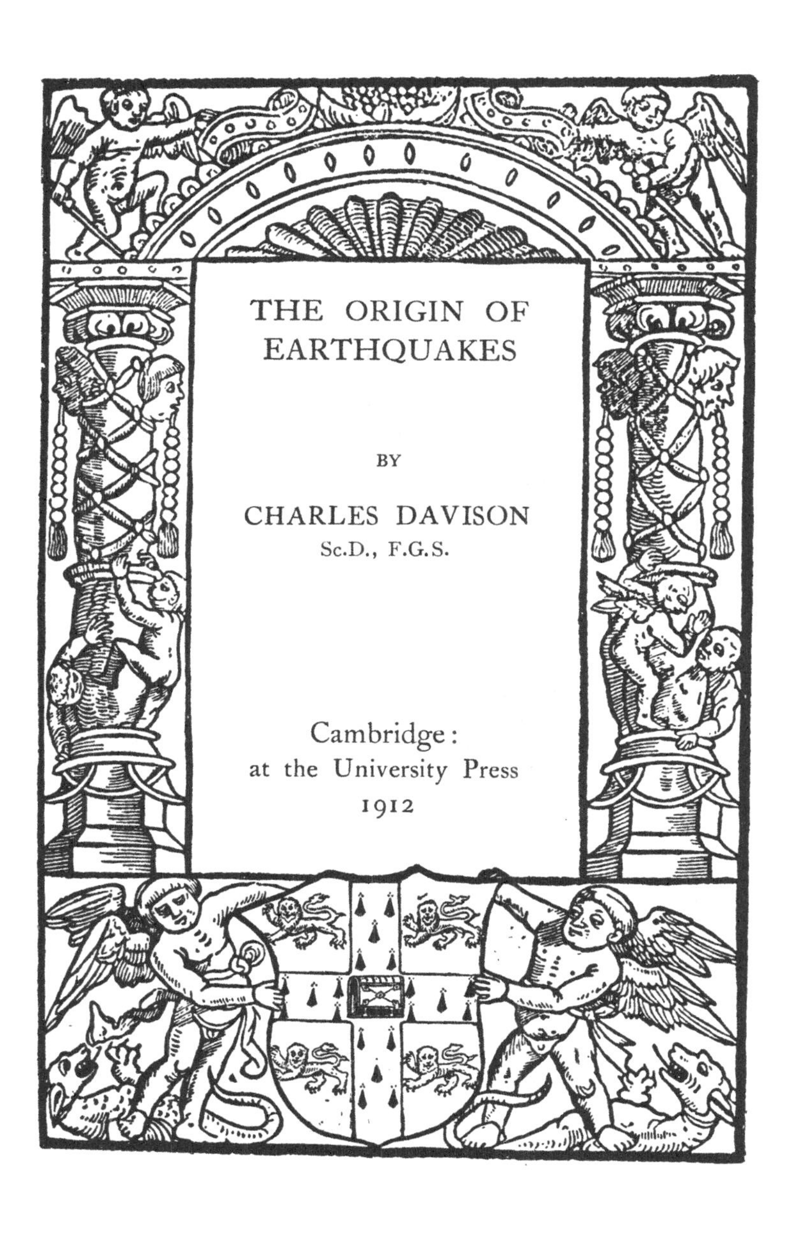

THE ORIGIN OF EARTHQUAKES

BY

CHARLES DAVISON
Sc.D., F.G.S.

Cambridge:
at the University Press
1912

CAMBRIDGE UNIVERSITY PRESS
Cambridge, New York, Melbourne, Madrid, Cape Town, Singapore,
São Paulo, Delhi

Cambridge University Press
The Edinburgh Building, Cambridge CB2 8RU, UK

Published in the United States of America by
Cambridge University Press, New York

www.cambridge.org
Information on this title: www.cambridge.org/9780521112215

First published 1912
This digitally printed version 2009

A catalogue record for this publication is available from the British Library

ISBN 978-0-521-04799-9 hardback
ISBN 978-0-521-11221-5 paperback

*With the exception of the coat of arms at
the foot, the design on the title page is a
reproduction of one used by the earliest known
Cambridge printer, John Siberch, 1521*

PREFACE

AN attempt to compress a great subject into a small compass involves the suppression of many details that would naturally have found a place in a larger volume. When the subject is one that has never yet been allotted more than a few pages, the difficulty of selection is increased. I can only trust that, in considering the origin of different classes of earthquakes and in describing fully one or two types of each, I may have succeeded in depicting the place of earthquakes in the realm of nature more satisfactorily than if I had filled these pages with a greater mass of detail. But I regret the omission of all reference to the remarkable fault-scarps of the Owen's Valley, Alaskan, Formosan and other earthquakes, to the periodicity of earthquakes, or to the almost simultaneous occurrence of great earthquakes in widely-separated portions of the globe.

The theory described in the following pages is one that has naturally occurred to many persons. So far as I know, the earliest reference to it is contained in a paper on the Visp earthquake of 1855 by the present 'Father' of English geologists, the Rev. Osmond Fisher.

CHARLES DAVISON.

February 1912

CONTENTS

LIST OF ILLUSTRATIONS

CHAPTER I

THE EARTHQUAKE PHENOMENA

IF we consider all the varied phenomena of earthquakes, we shall find that some are essential to, while others are of little consequence in, any explanation of their origin. Some are closely connected with the impulse which gives rise to an earthquake; others are merely the results of the passing shock. To draw the line between the two classes may not always be easy, but certain phenomena may at once be disregarded as having little, if any, bearing on the origin of earthquakes. The fissuring of the surface-soil on slopes and river-banks, the ejection of water from the ground, the starting of landslips from mountain-sides, these are all mechanical effects of the shock and may be left out of account in the present inquiry. This first chapter is therefore confined to a summary of the more essential phenomena of earthquakes, those which must be explained on any consistent theory of their origin.

1. An important feature of earthquakes is that they are not distributed uniformly over the globe, but are more frequent and more violent in certain districts. The earthquakes of one class, such as those felt on the flanks of Vesuvius or Etna or in the island of Ischia, are confined to the neighbourhood of active or dormant volcanoes. They are evidently connected with the operations which, after an interval more or less long, may result in a volcanic eruption. Others, and these are much the more important, originate in districts that are far removed from an active volcano. Such were the Lisbon earthquake of 1755, the Indian earthquakes of 1897 and 1905, the Californian earthquake of 1906, and the Messina earthquake of 1908.

As a rule, active volcanoes are singularly free from disastrous earthquakes. In Italy, the sides of Etna and Vesuvius are at times strongly shaken by local shocks, but the origins of the great Calabrian earthquakes are remote from any active vent. In Japan, where volcanoes are numerous, the districts surrounding the volcanoes are less frequently shaken than other parts of the islands. For every earthquake felt in the neighbourhood of a volcano—and all such are not necessarily of volcanic origin—five are felt on an average in equal areas elsewhere. Destructive earthquakes are practically confined to non-volcanic regions, and chiefly to the steeply sloping ocean-bed to the east of the Japanese empire. In the same way, the disastrous earthquakes of South America originate beneath the Pacific Ocean, in districts that are distant many miles from the volcanic vents of the Andean chain. The great earthquakes of India and Turkestan, again, occur in countries from which volcanic action is now entirely absent, but which lie on the steep sides of mountain-ranges that are known to be of recent growth.

2. A second feature, no less important and equally evident, is the extraordinary variation in the strength of earthquakes. On the one hand, are shocks like the Indian earthquakes of 1897 and 1905, which were felt over areas of nearly two million square miles, and in the latter case to a distance of a thousand miles from the centre. On the other hand, are the feeble shocks of this country, the majority of which disturb areas of much less than a hundred square miles and are insensible beyond a few miles from their origin. Still lower in the scale are earth-sounds, or noises unaccompanied by any perceptible shaking. But between the greatest of earthquakes and the weakest of tremors or earth-sounds, no boundary-line can be drawn, for, from one end of the scale to the other, every intermediate degree of strength is manifested.

This continuity in the variation of strength does not imply that the cause of all earthquakes, great and small, is the same. But it seems unnecessary to assign one cause for the tremendous disturbances of

India and central Asia, and to appeal to another for the slight movements of Norway and Great Britain, if a single cause can be indicated which is capable of acting with every gradation of strength from the extremity of violence to the last degree of feebleness.

3. A detailed map of an earthquake contains a series of curves, called *isoseismal lines*, each of which passes through all places at which the shock was of the same degree of strength. The outermost line of all bounds the *disturbed area* of the earthquake, and is that along which the shock can just be felt by the unaided senses.

In very slight shocks, the boundary of the disturbed area and the isoseismal lines are nearly circular in form; but, in earthquakes of moderate or great intensity, while the outer lines may still be nearly circular, the inner curves are distinctly elongated, the longer diameter of the curves being twice, or even three or more times, as long as the shorter diameter. The subject will be referred to in greater detail in the next chapter. For the present, it will be sufficient to notice the forms of the isoseismal lines represented in figs. 3, 7, 11 and 14.

4. In classifying non-volcanic earthquakes, it is convenient to take advantage of differences in the nature of the shock; but, in dividing such earthquakes into *simple*, *twin* and *complex*, it will be seen afterwards that, while all result from the growth

of the earth's crust, the classification nevertheless corresponds to important differences in their mode of origin.

In *simple* earthquakes, the shock begins with faint but rapid tremors, like those experienced during the passage of a heavy train or waggon, about four or five tremors being felt in every second.

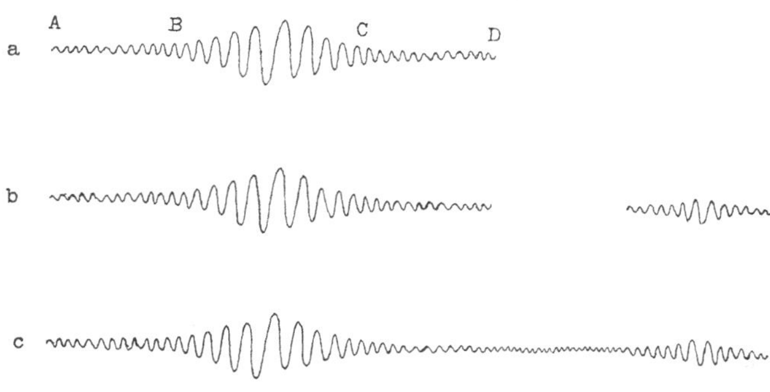

Fig. 1. Vibrations of simple and twin earthquakes.

After a brief interval, these tremors increase in strength. They merge rapidly into vibrations of greater range and duration, not more than two or three occurring in a second, while, in strong earthquakes, each vibration may last a second or more. In these vibrations, there may be variations in strength, but, as a rule, when once the maximum is reached, they decrease quickly in strength, and the shock, so far as it is sensible to the body, ends, as it began, with a weak tremulous motion. The whole movement may be represented graphically by the curve in fig. 1 *a*, in which the portion *AB* represents the preliminary tremors, the portion *BC* the principal vibrations, and the portion *CD* the concluding tremors. The total duration of simple earthquakes varies considerably. In Great Britain, they last about four or five seconds, occasionally as much as seven or eight. In other countries, with stronger earthquakes, the duration may be much greater.

In *twin* earthquakes, the movement is similar to that in simple earthquakes, but it is repeated after the lapse of two or three seconds, during which little or no motion is felt. The curves in fig. 1 *b* give a graphical representation of this peculiarity. In some parts of the disturbed area, a weak tremor may be felt during the interval between the main parts of the shock, as represented in fig. 1 *c*; but, at some distance from the centre, these tremors become imperceptible and the shock consists of two detached parts. The parts may differ in strength and duration; and these differences are not constant throughout the disturbed area, the first part being in some places the stronger, and in others the second. Twin earthquakes are necessarily of longer duration than simple earthquakes. In Great Britain, the total duration varies from seven or eight to about fifteen seconds. In strong twin earthquakes, as in the Charleston earthquake of 1886 or the Messina earthquake of 1908, the duration may be as much as a minute or even more.

Among *complex* earthquakes must be reckoned the greatest of all seismic disturbances. The shocks last much longer than those of the other classes, some for even three or four minutes. During this time, there are many variations in strength and rapid changes of direction. The ground is sometimes displaced so violently that persons are unable to stand. Waves are seen to travel along its surface and, as they pass, fissures open and close up again. In great complex earthquakes, the ground itself is rent and crumpled, and, along fractures which run persistently in one direction for many miles, the solid crust on one or both sides may be shifted horizontally as well as elevated or depressed.

5. The first symptom of a coming earthquake is generally a low rumbling sound, so low that, to many observers not otherwise deaf, it is quite inaudible. The sound grows rapidly louder, and with it the first tremors become perceptible. Both continue to increase in strength until the principal vibrations are felt, when deep explosive crashes are often heard in the midst of the rumbling sound. The sound and shock then die away together, the former often

continuing for a few seconds after the vibrations have become insensible.

The earthquake-sound will be considered more fully in a subsequent chapter. The principal phenomena which bear on the origin of earthquakes are, however, its great depth, its general precedence with regard to the shock, the small area over which the sound is heard in strong earthquakes, and the fact that the sound-area is not concentric with the isoseismal lines.

6. A strong earthquake sometimes, like the Messina earthquake of 1908, occurs without any warning. Occasionally, as in the case of the Charleston earthquake of 1886 or the Riviera earthquake of 1887, it is preceded by a few slight tremors, known as *fore-shocks*. Most earthquakes, and all strong earthquakes, are, however, followed by long trains of *after-shocks*. For the first few days after a great earthquake, the ground in the central district hardly ever comes to rest. The series of after-shocks may last for months or even years. Interspersed among them are occasional strong shocks, seldom comparable in intensity with the parent earthquake, but each followed by its own train of after-shocks.

7. While the after-shocks decline rapidly in frequency and as a rule in intensity, their centres shew a continual tendency to migrate from one region to another. Examples of this migration will

be given in succeeding chapters and also in Chapter VIII, which deals specially with after-shocks.

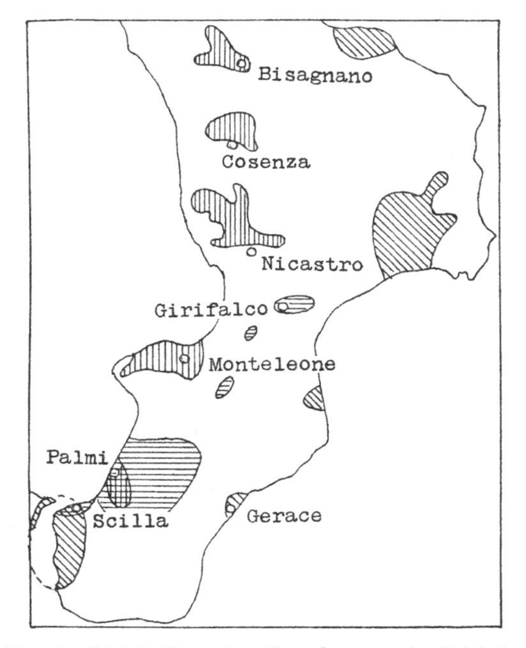

Fig. 2. Distribution of earthquake-zones in Calabria.

The same tendency to visit different districts in succession characterises the earthquakes of larger areas than that beneath which a single earthquake originates. The map in fig. 2, for instance, shews the distribution of earthquakes in Calabria and northern Sicily during the last century and a quarter. The small shaded areas are those most strongly shaken in different earthquakes, and in all probability are immediately over the *seismic foci* or regions in which the earthquakes originated. The areas shaded horizontally are those belonging to the remarkable series of Calabrian earthquakes in 1783 and the following years. Those shaded with vertical lines correspond to the earthquake of 1905, while those shaded obliquely are the central areas of various other earthquakes.

Sometimes, different centres are in action successively in one series of earthquakes. For instance, in 1783, the first great earthquake occurred on February 5 in the Palmi zone, the second a few hours afterwards in the Scilla zone. Two days later, the third great earthquake of the series took place in the Monteleone zone, followed in two hours by a fourth in the Messina and Scilla zones. On March 1, a fifth great earthquake occurred in the Monteleone zone, and on March 28, a sixth, almost, if not quite, as strong as the first of the series, in the Girifalco zone.

On other occasions, different centres come into action simultaneously or perhaps in very rapid succession. In the destructive earthquake of 1905, this was the case with the five centres corresponding to the Palmi, Monteleone, Nicastro, Cosenza and Bisagnano zones. At other times, single centres only are in action, as, for instance, that of Palmi in 1894, Monteleone in 1659, Nicastro in 1638, Cosenza in 1854, and Bisagnano in 1836. The Messina earthquake of 1908 originated in two detached centres situated beneath the Straits of Messina.

Now, the fact that these zones may be disturbed singly or successively in a series of earthquakes, seems to indicate that there are as many corresponding centres or foci. On the other hand, the fact that they may be disturbed simultaneously, or nearly so, shows that there must be some intimate connexion between them.

8. Lastly, one of the most significant features of earthquakes is their origin at a comparatively small depth below the surface. What this depth may be, we have no means of measuring with any approach to accuracy. Though many methods have been devised for its determination, all are faulty in principle or difficult to apply. They agree, however, in assigning in most cases a depth of only a few miles, and this is the one result of all the many estimates that have been made that can be regarded as of any value. We are, moreover, led to the same conclusion by another line of evidence. If earthquakes originated at a depth of many miles, say twenty or more, the intensity of the resulting shock would decline very

slowly with increasing distance from the centre of the disturbed area. Now, the area over which a slight shock is felt is invariably small, in many cases only a few miles in diameter. In a great earthquake, the chief destructive effects are also manifested within an area which, though it may be of considerable length, is limited in width (see, for instance, figs. 11 and 14). Thus, whether the shock be weak or strong, it is clear that as a general rule the centre is situated at not more than a few miles below the surface.

Whether it be situated at a depth of less than a mile or at one of several miles below the surface is of little consequence in our present inquiry. The important point is that it is confined for the most part to the superficial layer of the earth's crust. Now, if our knowledge of the nature of the earth's interior were confined to that of a skin less than a mile in depth, we might be compelled, in seeking for the origin of earthquakes, to assume the action of forces of the existence of which we have no direct evidence. But many of the rocks now at the surface were once at the depth of several miles ; and, from the structure of these rocks, we can infer the structure of the crust at a considerable depth. We have no reason for thinking that the forces which produced that structure in past geological periods have ceased to work. Rather must we suppose them to be acting still, with lessened efficiency perhaps, but in the same manner as of old and attended with similar results. Before we appeal to agencies of which we have no actual evidence, should we not therefore inquire whether the forces which have elevated our mountain-ranges and crushed the surface rocks into new and varied forms, can achieve their work without the accompaniment of great earthquakes and countless tremors? For no machinery, however perfect its construction may be, can move entirely without noise and friction.

CHAPTER II

EARTHQUAKES AND THE GROWTH OF FAULTS

THE movements which have brought the earth's crust into its present condition, which have elevated our mountain-ranges and depressed its borders below the level of the sea, take place in two principal ways, in the folding and crumpling of the rock, and in the fracturing of the crust and the subsequent displacements which convert the fractures into faults. Of these two modes of motion, folding, it is evident, must take place slowly and as a rule without violent interruptions. There are limits, however, to the extent to which the process may be carried, and folding, when pressed beyond these limits, merges into and gives place to faulting. On the other hand, fracturing must be more or less sudden in its action. Sudden, also, and coming as a relief to prolonged straining, must be any movements of the crust along the surface of a fracture. It is possible that some earthquakes may be connected with the operations of simple folding. It is more probable that they are caused by the abrupt and violent process of faulting. It is the principal object of this volume to show that the great majority of earthquakes are due to the intermittent growth of faults, that, when a displacement occurs at some depth, the friction generated by the sudden sliding of one huge rock-mass over and against the other must produce an intense jar in the solid crust around, a series of vibrations which, propagated outwards in all directions, gives rise at the surface to an earthquake-shock; and that, in those somewhat rare cases, in which the displacement is continued right up to the surface, the sudden spring of the displaced crust must complicate and increase the shock due to the grating of the sliding masses. In the present chapter, evidence will be brought forward to show that earthquakes are connected with fractures rather than with crust-folding, and with the process of faulting rather than with that of fracturing. In subsequent chapters, the modes according to which faults grow will be traced in greater detail, so far as light is thrown on them by the different classes of earthquakes described in the first chapter.

That earthquakes are associated in some way with the growth of faults is shown, in the first place, by the elongated form of the disturbed area and isoseismal lines of most earthquakes. On the maps of the Inverness earthquake of 1901 (fig. 3) and the Derby earthquake of 1903 (fig. 7) are drawn a series of isoseismal lines which pass through all places at which the shock just attained certain definite degrees of intensity. In both cases, it will be seen that the inner curves are elongated, while the outer curves continually approach to a rough circularity in form. In the Inverness earthquake, the dimensions of the innermost isoseismal line are 12 and 7 miles, in the Derby earthquake 16½ and 8½ miles. In many other British earthquakes, of all degrees of strength, the elongation of the innermost isoseismal lines is equally marked. In the two Wells earthquakes of 1893, the dimensions of the inner curves are 9½ and 5, and 11½ and 5, miles; in the Exmoor earthquake of 1894, they are 23 and 12 miles. In one of the Carlisle earthquakes of 1901, a moderately strong shock, the dimensions are 37 and 13 miles. The Carnarvon earthquake of 1903 and the Swansea earthquake of 1906 belong to the class of earthquakes which, for this country, may be termed strong. In the former, the dimensions of the innermost isoseismal line are 33½ and 15 miles, in the latter 26 and 14 miles.

Turning to the earthquakes of other lands, numerous instances of the same elongation might be given, especially from countries in which the mountain-ranges are of recent origin. It will be sufficient to quote a few. In the Neapolitan earthquake of 1857, the dimensions of the innermost isoseismal line are 64 and 23 miles; in the Verny (Turkestan) earthquake of 1887, 52 and 27 miles; in the Greek earthquake of April 20, 1894, 17 and 5 miles; in the Constantinople earthquake of 1894, 109 and 24 miles; in the Baluchistan earthquake of 1909, 57 and 8 miles.

Two explanations of the elongation of the disturbed area and isoseismal lines have been offered; one that the seismic focus, or region in which the earthquake originates, is of considerable length and parallel to their longer axes, the other that the vibrations are transmitted with greater ease in this direction. It is not difficult to argue in favour of the latter explanation. The longer axes are generally parallel to mountain-ranges, and the rocks along their axes may be more continuous, there may be fewer bounding surfaces to be crossed, than in the perpendicular direction. No one who has studied the manner in which the damage in a stricken town varies with the underlying rock can be blind to the fact that the intensity of the shock is influenced by the nature of the rock on which it is felt. Nevertheless, if the explanation were correct, the isoseismal lines of an earthquake should be approximately similar in form, and so also should be those of different earthquakes in the same region. Now, this is very far from being the case. In one earthquake, the innermost isoseismal line may be circular, the next extremely elongated (fig. 8). Or, as in the Inverness earthquakes of 1901 (fig. 4), the isoseismal lines of some shocks may be elongated and those of others circular. Or, even, as in the Pembroke earthquakes of 1892, the isoseismal lines of some earthquakes may be directed north and south, and of others east and west.

We are led therefore to the alternative explanation, that the seismic focus is of considerable length and is approximately parallel to the longer axes of the isoseismal lines. It is evident, in this case, that the inner isoseismal lines should adhere closely to the form of the focus and should therefore be the more elongated; while the outer lines, so far as they can be drawn, should approach more and more nearly to circularity. That this is so will be seen from the maps of the Inverness and Derby earthquakes in figs. 3, 7 and 8.

The significance of the elongated forms of the isoseismal lines becomes apparent when they are drawn on a physical or geological map. Their longer axes are then seen to be either parallel or perpendicular to the axes of mountain-chains, to the main lines of folding, or to the principal faults of the district. In Great Britain, this is the case with all important earthquakes. In most of the Swiss earthquakes, the longer axes are parallel to the neighbouring chains of the Alps and Juras, though in a few instances they are in the perpendicular direction. The axes of the Greek earthquakes of April 20 and 27, 1894, are parallel to the neighbouring depression of the Gulf of Euboea. That of the Constantinople earthquake of 1894 coincides with the line of depression which begins at Ada-Bazar and is marked by the Lake of Sabandja and the Gulf of Ismid. The central area of the Tokyo earthquake of 1894 is a north and south band which occupies the lowest part of the plain of Musashi, the plain itself being the continuation of the axis of the Bay of Tokyo. In the great Japanese earthquake of 1891 and the Californian earthquake of 1906, the areas most strongly shaken are narrow bands following the courses of the remarkable faults that will be described in later chapters (figs. 11 and 14).

We are thus led again to the conclusion that earthquakes are closely related to one or both of the two processes of folding and faulting to which the outlining of the earth's great surface-features is chiefly due. But the fact that, in any district, the axes of the disturbed areas and isoseismal lines are parallel, not to one line only, but to two lines roughly at right angles to one another, indicates that they are not directly connected with the great folds and mountain-axes, but rather with the double system of faults which, it is well-known, have a tendency to cross at right angles, one series (called *strike* faults) being approximately parallel, and the other (called *transverse* faults) perpendicular, to the main lines of folding.

Again, when successive earthquakes in a district are mapped, the central areas are sometimes found to be nearly coincident; at other times, the centre is displaced, and the axes of the new isoseismal lines may then be either parallel or perpendicular to those of the old ones or else in the same straight line with them. Some examples of this migration of seismic foci along a fault will be found in the third and fourth chapters, on simple and twin earthquakes (figs. 4 and 5).

Lastly, as will be described in Chapters V—VII, a few violent earthquakes have certainly been accompanied by the formation of fault-scarps; and, in these cases, there can be no doubt as to the connexion between the two phenomena.

The aim of the preceding paragraphs is to prove the reality of the connexion between earthquakes and systems of faults. Now, earthquakes may be produced during fault-formation in two ways, either by the mere act of fracturing or by the friction generated by slipping. Possibly slipping may take place immediately after fracturing, so that a single earthquake might be due in part to both causes. Some reasons, however, will now be given for concluding that fault-slipping, rather than fracturing, is responsible for the great majority of earthquakes.

Firstly, during the actual fracturing which initiates a fault, very few earthquakes could be produced, one by the first fracture and a few more by its further extension. But the subsequent displacement must be the result of innumerable slips. Now, the number of earthquakes originating in a given district is enormously in excess of the number of faults in it, and the inequality will be the more evident if we consider that in many districts faults have been in process of formation during a large part of geological time, while our earthquake-records extend over at the most a few centuries, often over only a few years. For instance, 143 shocks and earth-sounds were noted at Comrie, in Perthshire, during the last three months of 1839 ; 306 shocks were felt in the island of Zante during the year 1896 ; while at Gifu, in Japan, 3365 shocks were recorded between October 28, 1891, and the end of the year 1893. There are reasons for connecting the earthquakes at all three places with neighbouring faults, and it is quite inconceivable that each shock was caused by the formation of a new fracture or the extension of an old one.

Again, in several cases, though not in many, it is certain that fault-slips have taken place with earthquakes, for they have extended to the surface and been left visible there as fault-scarps. During the great Mino-Owari (Japan) earthquake of 1891, as will be described in Chapter VI, an extraordinary fault-scarp was formed. It was actually traced, cutting through hills and across plains, for a distance of 40 miles, and its total length was probably not less than 70 miles. In one part, the vertical displacement reached 18 or 20 feet. Moreover, the distribution of earthquake-centres in the same region during the two previous years (fig. 19) shows that the fault-system (for there may have been more than one fault in action) was a seat of rather frequent shocks, so that it is improbable that the great earthquake was due to the formation of a new fracture. Other examples of fault-scarps will be described in Chapters V and VII. The number known to us is small compared with the total number of recorded earthquakes, but, in all probability, some are also formed in the bed of the ocean and, by the sudden translation of a large body of water above, give rise to sea-waves which are propagated from one side of an ocean to the other.

Thirdly, the migration of the seismic focus along a fault has already been referred to. If it were to take place outwards in one or both directions only, this migration might be due to an enlargement of the fracture; but, since the foci generally retrace their steps and even oscillate from side to side, the resulting shocks must rather be referred to fault-slipping.

Lastly, it is obviously less difficult to generate or precipitate a fault-slip than to make a new fracture; and this is a matter of not a little consequence if we reflect that, in all probability, faults are ultimately due to the slow cooling of the earth. Moreover, in the case of fracturing, the initial disturbance is merely the elastic recoil of the rock-particles from the surface of fracture; whereas, in the case of slipping, the original disturbance depends on the weight of the displaced crust. As this weight may vary from that of a few cubic miles to one of several thousand cubic miles, and the displacement from a fraction of an inch in the one case to several yards in the other, it is evident that the friction so generated must be capable of producing earthquakes as slight as those which are felt at Comrie or as disastrous as those which visit the coasts of Chili and Japan.

CHAPTER III

SIMPLE EARTHQUAKES AND THEIR ORIGIN

THE characteristic feature of a simple earthquake is that the shock should consist of a single series of vibrations, as a rule with only one maximum of intensity. Of the earthquakes felt in this country, the great majority—about 95 per cent.—are simple

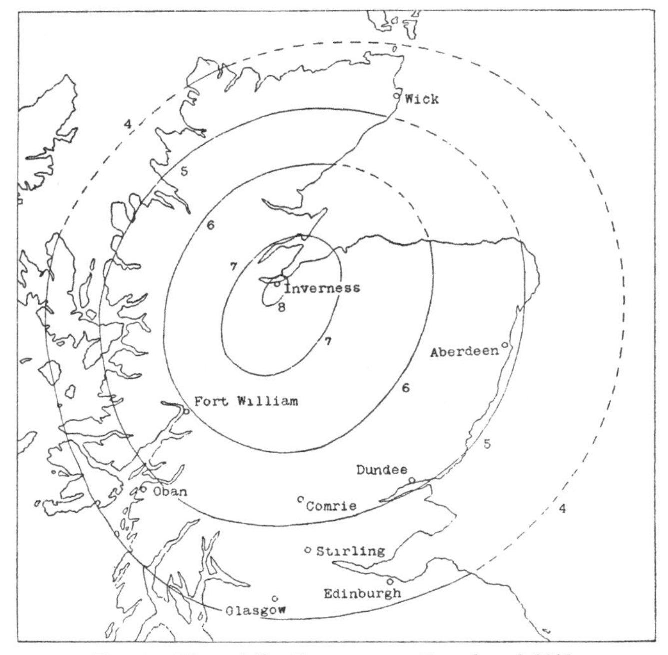

Fig. 3. Map of the Inverness earthquake of 1901.

earthquakes. A few of them, like the Inverness earthquake of 1901, described in the present chapter, are of considerable intensity. The greater number are weak tremors and disturb areas of less than one or two hundred square miles.

The accompanying map (fig. 3) shows the area disturbed by the Inverness earthquake of September 18, 1901, and a series of isoseismal lines, which illustrate the manner in which the intensity of the shock varied throughout that area. Within the isoseismal line marked 8, the shock was strong enough to cause slight damage to property; chimney-pots in many houses were thrown down and walls were cracked. Outside this line and within the curve marked 7, the shock failed to cause damage but was strong enough to throw down ornaments, vases, etc. In the next zone, that bounded by the isoseismal line of intensity 6, the shock made chandeliers, pictures and other suspended objects swing. Outside this, in a fourth zone bounded by the isoseismal line of intensity 5, the shocks caused the observer's seat to be perceptibly raised or moved. In the last zone, bounded by the isoseismal 4, the shock was strong enough to make doors, windows or fire-irons rattle. Farther still, the shock does not seem to have been felt by any person, and this last line may therefore be regarded as the boundary of the disturbed area, which contains about 30,000 square miles.

The chief peculiarity of these isoseismal lines is that, while the two outer ones are nearly circular, the three inner curves are distinctly elongated, and all in the same direction, roughly northeast and southwest, more accurately N. 33° E. and S. 33° W. Another point, to which reference will be made afterwards, is that the successive curves are farther apart on the southeast than on the northwest side.

Besides this strong shock, at least one minor shock was felt shortly before, and fifteen within the two months following it. The boundaries of the areas disturbed by some of them are shown in fig. 4. Most of them are elongated in the same direction as the isoseismal lines of the principal earthquake, but one at least is approximately circular. This exception shews that the distortion of the curves in the northeast and southwest direction is due, not to any peculiarity of the surface-rocks by which the vibrations are transmitted with greater ease in this line than in any other, but rather to the fact that the seismic focus was itself elongated in this direction.

How strongly this direction is impressed on the geography of the central district will be evident at once from a glance at the map (fig. 3). Beginning at the northeast, there is the long straight southeast coast of Ross-shire extending for a distance of thirty miles, from Tarbat Ness, pass Cromarty and Fortrose to the neighbourhood of Inverness. The line of coast is continued in the same direction southwest-wards by the series of depressions connected by the

Caledonian canal, and occupied by the lochs of Dochfour, Ness, Oich, and Lochy, ending with the sea-lochs of Eil and Linnhe.

This striking linearity of structure is known to be due to the presence of the Great Glen fault, which traverses the whole of Scotland along the line indicated. Roughly, the fracture may be regarded as nearly a plane surface, inclined towards the southeast, along which the rocks on that side have been thrown downward through several hundred feet. Partly perhaps by unequal depression along the line, partly by the exposure of rocks differing in hardness and power of resistance to weathering, the formation of the fault has left its mark upon the district.

That the earthquakes of 1901 were caused by a series of small belated movements along this great fault is rendered probable by the distortion of the isoseismal lines in the direction of the fault-line (fig. 3) as well as of the boundaries of the areas shaken by the more important minor shocks (fig. 4). There is, however, other and more conclusive evidence of the connexion suggested. The seismic focus being approximately a plane surface inclined to the horizon, the intensity of the shock must be greatest on the side towards which the fault-surface is inclined, and must fade away more slowly on that side than in the opposite direction. In the neighbourhood of the centre, the isoseismal lines should thus be farther

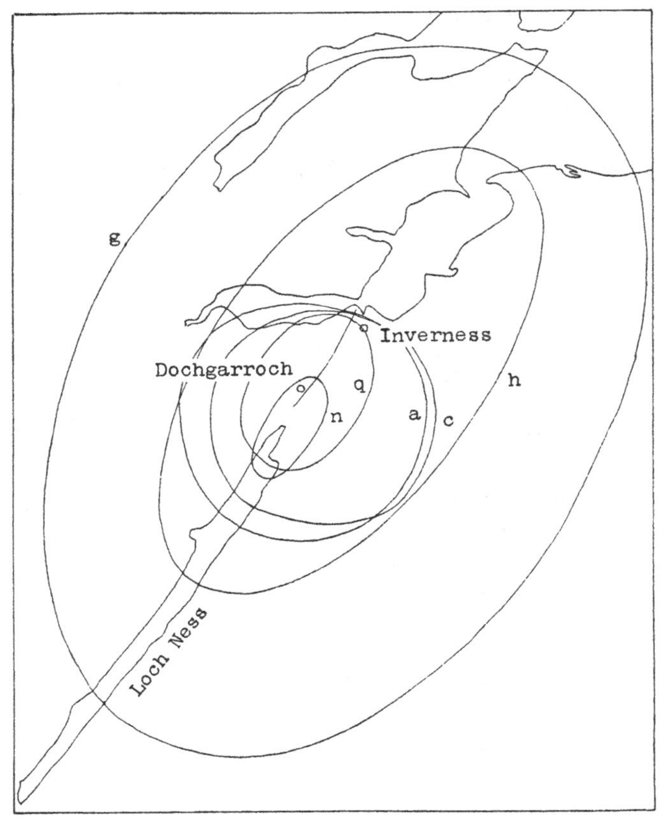

Fig. 4. Map of the principal after-shocks of the Inverness earthquake of 1901.

apart on the side towards which the fault is inclined than on the other, and the centre of the innermost isoseismal line should also be on that side.

If, then, the Inverness earthquakes were due to fault-slipping, the fault must run in a direction from about N. 33° E. to S. 33° W., the fault-surface must be inclined to the southeast, and the fault-line must lie a short distance (a mile or more) on the northwest side of the centre of the isoseismal line of intensity 8.

Now, the mean direction of the Great Glen fault near Inverness is N. 35° E. and S. 35° W., it is inclined to the southeast, and the course of the fault-line (as shown in fig. 4) is a short distance to the northwest of the centre of the isoseismal 8. The correspondence between the known fault and that indicated by the evidence of the isoseismal lines is so close that little doubt can be felt as to the earthquake being caused by a slip along this well-known fault.

If further evidence as to the connexion were needed, it is provided by the distribution of the minor shocks which preceded and followed the principal earthquake. The boundaries of the areas shaken by the more important shocks are shown in fig. 4, and the centres of these areas in fig. 5. There was first a slight shock (*a*) at 6.4 p.m. on September 16. Then came the principal earthquake at 1.24 a.m. on September 18, the focus of which was several miles in length and extended nearly from Loch Ness to

Inverness. Ten minutes later, a slight after-shock (*c*) occurred near the southwest margin of the principal focus. After 2¼ hours, the chief after-shock (*g*) took place. Its centre was half a mile farther to the north-east, but, as its focus was several miles in length, it

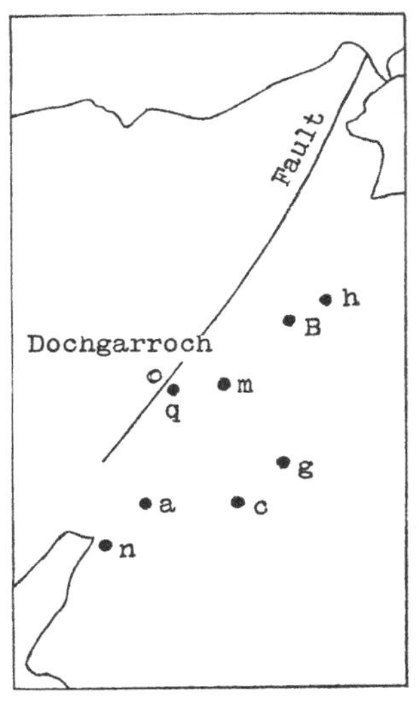

Fig. 5. Distribution of the centres of the principal after-shocks of the Inverness earthquake of 1901.

must have extended some distance beyond the south-west margin of the principal focus. Five hours afterwards, another shock (*h*) occurred, its centre was about half a mile northeast of the principal centre, and its focus probably reached beyond, but not much beyond, the northeast margin of the principal focus. During the next eleven days there were no movements of any importance, but, on September 29, a small slip (*m*) occurred about one mile to the southwest of the principal centre (*B*). This was followed, on September 30, by one of the principal after-shocks (*n*), the centre of which lay to the southwest of the principal focus, and the focus of which must have extended two or three miles beneath Loch Ness. Again, after about a fortnight more, there was a shock (*q*) on October 13, the focus of which was about two miles long in the neighbourhood of Dochgarroch. Thus, the foci of all the after-shocks, as well as the focus of the principal earthquake, lie within a narrow band parallel to the fault and on the southeast side of it, that is, on the side towards which the fault-surface is inclined. It will be noticed, also, that in the later shocks there is a gradual approach of the centre towards the fault-line, indicating that the depth of the corresponding foci gradually decreased with the lapse of time.

With regard to the nature of the slips which gave rise to the Inverness earthquakes, we have no evidence. We do not know on which side of the fault the rock moved, or whether it moved on both sides, though unequally, at the same time. In past ages, however, the chief movement seems to have been one of the southeast side downwards, forming the depressions of the Moray Firth and the chain of lakes. If the latest movements were of the same nature, we may imagine that the principal earthquake was caused by a slight but sudden sag of the rock on the southeast side over a length of about eight miles from Inverness to near Loch Ness.

In the accompanying diagram (fig. 6), the upper line is supposed to represent the surface of the earth,

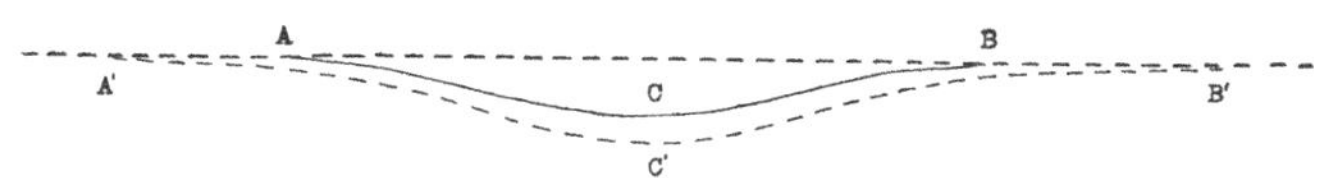

Fig. 6. Nature of the displacement causing a simple earthquake.

from the Moray Firth (northeast of Inverness) to Loch Ness. The lower dotted line is intended to represent a horizontal straight line traced on the south side of the fault-surface through the focus of the principal earthquake. After the movement which resulted in the earthquake, this line, in consequence of the sagging of the rock, takes up a curved position represented by the curved line ACB. The distance AB in this case represents the length of the focus, about eight miles. The distance between the dotted and curved line at *C* represents the amount of the subsidence. This distance is, for clearness, greatly exaggerated in the diagram. In reality, it was possibly not more than a fraction of an inch.

The effect of this displacement would in the first place be to increase the strain along the fault-surface in the terminal regions, *A* and *B*, of the principal focus. If the rock were previously on the point of slipping, or nearly so, the additional strain would probably overcome the lingering resistance, and we should therefore expect further slips to take place in these regions. The effect of these slips, again, would be to cause slightly increased strains still farther outwards and again in the central region below *C*, and this would continue, the slips becoming less and less in amount until the additional strains imposed could no longer overcome the resistance to movement. The final form of the straight line *AB* would thus be represented by the broken line *A′C′B′*.

It is evident, also, that a downward movement at the level of the focus would throw an additional strain on the fault-surface in the region above. In other words, there would be a tendency for the foci to diminish gradually in depth.

Now, of the six principal after-shocks of the Inverness earthquake, the first two occurred at and slightly beyond its southwest margin, the third at and beyond the northeast margin, the fourth near the centre and nearer the surface, the fifth beyond the southwest margin and at a still smaller depth, while the last was not far from the centre and quite close to the surface.

The effect of the movements which caused the stronger after-shocks was thus to extend the focus of the principal earthquake in both directions, but especially towards the southwest. This conclusion is also supported by the distribution of the slighter after-shocks, the majority of which occurred in the central and southwest regions, and several of which undoubtedly originated beneath Loch Ness.

Our knowledge of previous Inverness earthquakes is less complete than in the case of those of 1901. So far as it goes, however, it leads to the same results. The movements which caused the earthquakes of 1816, 1888 and 1890 took place beneath the neck of land between Inverness and Loch Ness, and in the first and last cases were followed by minor movements farther to the southwest, a few reaching the region beneath Loch Ness.

The apparent tendency of these various movements, small though they may be now, is thus to deepen the east end of Loch Ness, and so to provide for its gradual extension northeastwards, until finally, after the lapse of many thousand years, Loch Ness and Loch Dochfour may coalesce and together form one long arm of the sea.

The Inverness earthquakes do not differ materially from other simple earthquakes. The attendance of minor shocks is not an invariable feature. Corresponding to the single series of vibrations, increasing to a maximum and then dying away, there is in each case a single focus in which the displacement increases to a maximum near its centre and then decreases to zero.

A distinctive property of simple earthquakes is that they are almost always connected with strike or longitudinal faults or faults that are parallel to the main lines of crust-folding; while twin earthquakes, as will be seen in the next chapter, are related to transverse faults or faults which cross the crust-folds obliquely or more or less nearly at right angles. The after-shocks of twin earthquakes are generally simple earthquakes; but, when simple earthquakes occur alone or in series, they represent the normal mode of growth of strike faults.

CHAPTER IV

TWIN EARTHQUAKES AND THEIR ORIGIN

If we may judge from the earthquakes of our own country, the vast majority, probably as many as 95 out of every hundred, of all earthquakes belong to the class of simple earthquakes. Most of the remainder are twin earthquakes. Complex earthquakes are few in number, perhaps sixty occur every year, though in their total effects they may exceed all the others together.

All the strongest earthquakes that are felt in this country are twins. The four greatest within the last quarter of a century disturbed areas (including the portions covered by the sea) of 98,000, 66,700, 63,360 and 44,860 square miles respectively, all of which except the last are larger than the area of England. Those here taken as the types of twin earthquakes, the Derby earthquakes of March 24, 1903, and July 3, 1904, were felt over areas of 12,000 and 25,000 square miles. Of the two, however, the former was the stronger. It was felt over a smaller area as it occurred on a weekday afternoon at half-past one. The other took place on a Sunday afternoon, at 3.21, and was noticed at great distances by many observers who were resting at the time.

The area disturbed by the earthquake of 1903 is shown in fig. 7. The curves are isoseismal lines corresponding to the same degrees of intensity as those in the map of the Inverness earthquake (fig. 3), except that there is no curve of intensity 8. A few houses here and there within the central area were slightly damaged, but the houses so injured were old and poorly built. There was no general overthrow of chimneys or fissuring of house-walls.

As in the Inverness earthquake, it will be noticed that the inner isoseismals are elongated in form and all in the same direction, namely, from N. 33° E. to

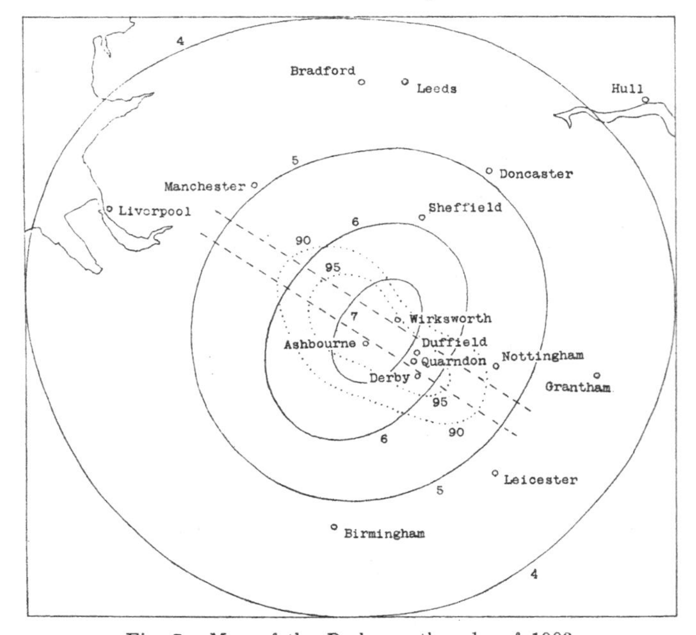

Fig. 7. Map of the Derby earthquake of 1903.

S. 33° W., and that the isoseismal lines are farther apart on the northwest than on the southeast side. We infer from this that the average direction of the fault which gave rise to the earthquake must be about N. 33° E. and S. 33° W. and that the fault-surface must slope to the northwest.

In the disposition of the isoseismal lines of the Derby earthquake of 1903, there is nothing to distinguish the earthquake from an ordinary simple earthquake such as that of Inverness in 1901. In the nature of the shock, however, there was a marked difference. In nearly all parts of the area illustrated in fig. 7, the shock consisted of two parts separated by a very brief interval of rest and quiet. The following accounts are given to illustrate the twin character of the shock and its mode of variation throughout the disturbed area. Of the places mentioned, Ashbourne and Darley Dale are close to the longer axis of the inner isoseismal lines, and respectively south and north of their centre; Duffield lies along the continuation of the shorter axis of the same curves, while Quarndon is 1½ miles, and Derby 3 miles, from the shorter axis.

At Ashbourne, the shock consisted of two distinct parts, the first being rather the stronger and lasting twice as long as the second; it seemed as if a heavy piece of furniture were rapidly rolled in the room upstairs from east to west, and then, after a pause of a second or two, were rolled a short way back again. At Darley Dale, there were also two parts, each of which began with a low distant rumbling like the rushing of a strong wind, and culminated in a violent shock as it seemed to pass beneath the house. In this case, the second part was the stronger.

At Derby, the shock also consisted of two distinct parts, each lasting three seconds and with an interval of half a second between them. At Quarndon, a rumbling sound was first heard; then came a violent shock, the rumbling continued for about two seconds, and, before it ceased, a second shock was felt, less violent than the first, after which the rumbling gradually died away. At Duffield, only a single shock was observed, a quivering motion during the loudest part of the rumbling sound, which resembled that made by a muffled peal of thunder or by a sudden gust of wind.

It is clear, from these accounts, that the nature of the shock varied throughout the disturbed area. In some parts, two distinct shocks were felt; in others, only one was observed. The law of variation is easily determined. Plotting on a map all the places where one shock or two shocks were noticed, the former are confined within a narrow straight band, about five miles wide, running centrally across the inner isoseismal lines in a direction from W. 34° N. to E. 34° S., that is, at right angles to the longer axes of the isoseismal lines. The boundaries of this band are represented by broken lines on the map in fig. 7. Outside the band, and throughout all the rest of the disturbed area, the interval between the two parts of the shock was one of rest and quiet, its average duration being three seconds. Close to the band (as at Derby) the interval was much shorter, though still distinct. Within the band, the two parts overlapped and the shock appeared single, with two maxima of intensity at places (like Quarndon) near the boundaries, with a single maximum at places (like Duffield) midway between them.

That the two parts of the shock were nearly equal in strength, is shewn from the facts that both parts were observed as far as the boundary of the disturbed area, and that, at any given place, observers differed in their estimates of the order of intensity of the two parts. At Derby, for instance, the first part of the shock was regarded as the stronger by 19 observers and the second by 16, while five considered the two parts as of approximately equal intensity. If we divide the disturbed area into two parts by the shorter axis of the isoseismal lines, three out of every five observers on each side of this line regarded the first part as stronger than the second.

To what are we to attribute the duplication of the shock throughout the whole of the disturbed area, with the exception of a narrow central band? A phenomenon so generally observed as the double shock cannot be explained by local peculiarities. We cannot, for instance, imagine that the shock was originally single and was apparently doubled by reflexion, so as to form an underground echo. Nor could the double shock be due to the propagation of two series of vibrations from the same focus, for, in that case, the first or the second part would everywhere be the stronger, and no explanation could be given of the existence of the band within which the single shock was observed. The facts described above can only be accounted for by the supposition that the shock originated in two distinct foci, within which occurred impulses of nearly equal strength. The exact positions of the two foci cannot be determined, but it is probable, from the form of the isoseismal curves, that one focus was situated roughly beneath Ashbourne, and the other beneath a place about three miles west of Wirksworth, their centres being thus about eight or nine miles apart.

Now, if the impulses within the two foci took place simultaneously, it is evident that the vibrations from both foci would coalesce and form a single shock along a straight narrow band traversing the disturbed area midway between the two foci and at right angles to the line joining them. If, however, the Wirksworth focus were the first in action by a second or two, the band within which the two shocks coalesced would be curved, with the concavity turned towards the southwest, for the vibrations from the Wirksworth focus would travel farther than those from the Ashbourne focus before the two series coalesced. Lastly, if the impulse at the Wirksworth focus preceded that at the Ashbourne focus by several seconds, the vibrations from the former would be felt first all over the disturbed area and thus no band would exist in which the two parts of the shock were felt simultaneously. This was the case in some twin earthquakes, such as the Doncaster earthquake of 1905 and the Swansea earthquake of 1906. In the Derby earthquake of 1903, it is clear that the two impulses occurred in the two detached foci at absolutely the same instant of time.

This conclusion is supported by the study of the sound-phenomena. The sound which accompanies an earthquake is a low rumbling, so low that, unless it be very loud, it may seem to be more felt than heard. To some persons, it appears as loud as if several traction-engines were passing; to others in the same place, and even in the same house, it passes entirely unheard. The latter are deaf, not so much because the sound was not loud enough, as because the vibrations were too slow to produce in their ears the sensation of sound. At places near the origin, the sound is, however, so loud that it is heard practically by all observers of the earthquake, but, as it declines in strength with increasing distance from the origin, it becomes inaudible to a larger and larger percentage of observers. Within the innermost isoseismal line, for instance, the sound was heard by 97 per cent. of the observers, and, in the zones included between successive pairs of isoseismal lines by 88, 80 and 64 per cent. respectively. But, though there is a general decline in audibility outwards from the origin, the rate of decline is not the same in all directions. This is shown by the two dotted curves on the map, the so-called *isacoustic* lines or lines of equal sound-audibility. The inner curve, marked 95, is such that if with any point on the curve as centre, a small circle were described, then, of all observers in the included area, 95 per cent. heard the sound, while to 5 per cent. it was inaudible. It will be noticed that the isacoustic lines are both distorted in the direction of the band within which the two parts of the shock coalesced. The reason is that the sound-vibrations from the two foci were heard simultaneously along and near this band, that they were therefore louder and heard by a larger percentage of observers in this district than elsewhere.

About forty days after the earthquake described above, on May 3, there occurred in the same district a simple earthquake of comparatively slight intensity, the total area disturbed by it being only about 585 square miles. The isoseismal lines are approximately parallel to those of the principal earthquake, and the centre of the innermost curve is about midway between the centres of the two foci of the principal earthquake. This after-shock must therefore have originated along the same fault as the principal earthquake, and in the region between the two detached foci of that earthquake.

For about fourteen months, there seems to have been no further movement in the Derbyshire district, until July 3, 1904, when another strong twin earthquake occurred. The area shown in fig. 8 is the same as that illustrated in fig. 7, and the isoseismal lines correspond to degrees of the same intensity. The curves, it will be noticed, are all of smaller dimensions than those of the earthquake of 1903, but most of the outer isoseismal line (that corresponding to intensity 3) is omitted. The innermost isoseismal line in this case is a small circle with its centre close to Ashbourne. The other curves are elongated, in almost exactly the same direction as in 1903, that is, from N. 31° E. to S. 31° W., and are farther apart on the northwest, than on the southeast, side. We infer that the fault runs in the direction indicated and that it is inclined towards the northwest. It must evidently therefore coincide with the fault which gave rise to the earthquake of 1903.

The shock again consisted of two parts, but the separation of the parts was not everywhere so distinct as in the previous year. Close to the centre, a weak tremulous motion was felt between, and connecting, the two parts. At some distance from the origin, this was imperceptible and the parts of the shock were completely separated, except along a narrow band within which the two parts were superposed

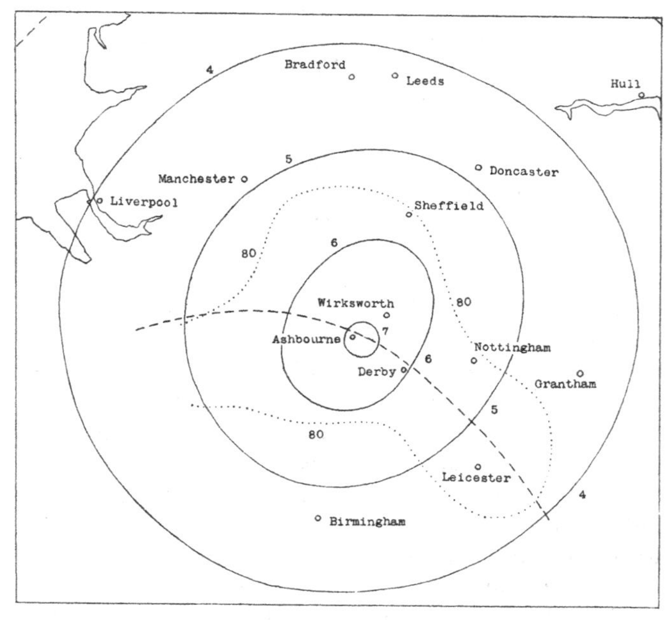

Fig. 8. Map of the Derby earthquake of 1904.

and appeared to form a single shock. The boundaries of this band cannot be determined with accuracy, but the central line of the band is indicated by the broken line in fig. 8. It will be noticed that this line is curved, its concavity facing the southwest, and that it crosses the longer axes of the isoseismal lines a short distance on the northeast side of the Ashbourne centre. We infer in this case, also, that the shock was caused by impulses in two detached foci, one near Ashbourne, the other to the west of Wirksworth, that is, in the same two foci that were responsible for the earthquake of 1903.

The two earthquakes did not, however, correspond in all respects. In 1903, the impulses in the two foci occurred simultaneously and were very nearly equal in strength. In 1904, the northeast or Wirksworth focus was in action first by a second or two. This is shown by the curvature towards the southwest of the band within which the two parts of the shock coalesced, for the waves from the northeast focus must have travelled farther before they were met by those from the other focus. Again, the two impulses, though nearly equal in strength, were not quite equal. The impulse within the southwest focus must have been the stronger, for this focus alone is surrounded by an isoseismal line of intensity 7. Moreover, the weaker part of the shock was felt over an area of only about 8000 square miles. Possibly, these differences between the two earthquakes may be due to the fact that, about an hour before the earthquake of 1904, a slight shock was felt in the neighbourhood of the northeast or Wirksworth focus. The small movement which caused it may have facilitated movement in the northeast focus, and at the same time reduced its intensity.

In the earthquake of 1904, the construction of isacoustic lines is less easy than in the case of its predecessor. Only one, represented by the dotted line marked 80, is shown, and that incompletely, in the map in fig. 8. Its form is less regular than in the corresponding curves for 1903, but it shows a marked distortion to the southeast along the axis of the band within which the single shock was observed. Such as it is, it confirms the conclusions that the shock originated in two detached foci and that the impulse in the northeast focus occurred a second or two before that in the southwest focus.

It is interesting to notice that the twin earthquake of 1904 was also followed, but this time after an interval of only eight hours, by a slight simple earthquake. The area disturbed by it was only 425 square miles. The isoseismal lines are again nearly parallel to those of the principal earthquake which surrounds both foci, the centre of the innermost isoseismal being about midway between the two centres of the twin earthquake. The after-shock of 1904, as in 1903, thus originated in the same fault as the other earthquakes and in the portion of it between the two foci which gave rise to the twin earthquake.

The most important feature of a twin earthquake is its initiation in two distinct foci. But equally significant, from the point of view of its origin, is the fact that in some twin earthquakes, such as those described above, the impulses take place either simultaneously in the detached foci or so closely together that one cannot be the consequence of the other. It may easily be imagined that the rock along a fault-surface might be on the point of slipping in two separate but neighbouring regions. If so, the waves resulting from a sudden movement in one region might precipitate a similar movement in the other. But, in these twin earthquakes, when the impulses do not occur simultaneously, it is evident that the second impulse takes place before the waves from the first focus have time to reach the second, otherwise the waves from the two foci could not coalesce in any part of the disturbed area. And when, as in the Derby earthquakes, similar twin-movements, separated by little more than a year, take place in the same twin foci, it is clear that some cause must be at work that is capable of producing simultaneous movements in these two detached regions.

In the case of a simple earthquake, the displacement which produces it is probably of the simplest possible character, a mere translation of the moving rock, so that all particles of the sliding rock-surface move along nearly parallel lines. In a twin earthquake, the only method by which movements can be effected simultaneously in two detached regions with little if any movement in the intermediate region is one of rotation about the latter region.

We may suppose the continuous line in fig. 9 to represent a section of a great fold in the earth's crust along a fault at right angles to its axis; *A* representing the crest of the fold or anticline, *S* the trough or syncline and *M* the median limb. Now, if a small step were to take place in the growth of the fold, from the form represented by the continuous line to that represented by the broken line, *A′MS′*, it is evident that there would be two regions of displacement, one between *A* and *A′*, and the other between *S* and *S′*, while, in the intermediate region about the median limb *M*, there would be little or no displacement. The two displaced regions would thus be the two seismic foci of the twin earthquake, more or less completely detached owing to the almost imperceptible movement of the median limb.

The theory of rotation of the median limb of a crust-fold thus accounts for the two detached foci and also for the simultaneous or nearly simultaneous movements within them, and it is difficult, if not impossible, to account for them in any other way. But, if the theory be correct, the distance between the foci should be approximately the same as that between successive crests and troughs of crust-folds.

Both distances are subject to wide variations. The distance between the twin foci may be as much as 23 miles, as in the Carlisle earthquake of 1901, or as little as 4 miles, as in the Colchester earthquake of 1884. The average distance in seven British twin earthquakes is, however, about 10 or 11 miles. For the lengths of British crust-folds, we have no detailed

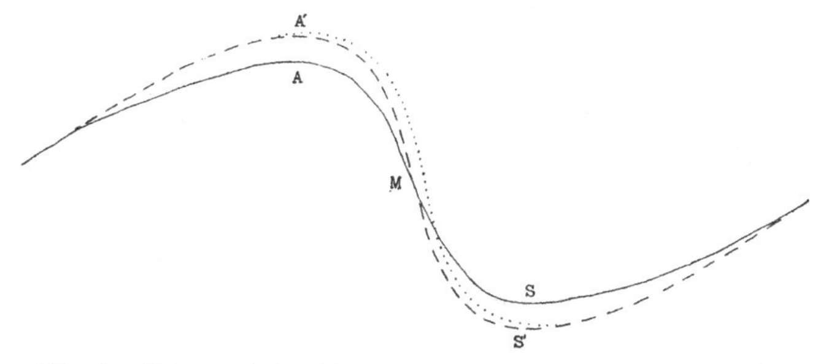

Fig. 9. Nature of the displacement causing a twin earthquake.

measurements, but the courses of the principal anticlines in France have been laid down, and the average distance between successive anticlines and synclines along several lines varies between 9 and 12 miles, which agrees closely with the average distance between the foci of the twin earthquakes.

Thus, while simple earthquakes are due for the most part to slips along strike faults, it follows that twin earthquakes are due to slips along transverse faults, that they are caused by the growth of crust-folds along fractures which intersect them transversely.

A movement of rotation of a crust-fold, such as that described above, must produce an increase in the strains already existing in the median limb. As this part of the fold is only just kept at rest during the movement which caused the twin earthquake, the sudden access of strain must cause a simple displacement of the median limb, into a position such as that indicated by the dotted line in fig. 9. Thus, within a longer or shorter interval after the occurrence of the twin earthquake, there should be a simple earthquake in the region of the fault intermediate between the foci of the twin earthquakes. This was the case with the two Derby earthquakes described in the present chapter, and it is indeed the usual result of a twin earthquake.

After-shocks, such as these, are invariably slight compared with the twin earthquakes themselves. They shew that the after-displacement of the median limb is small compared with that which takes place in the crest and trough of the fold. In other words, the occurrence of twin earthquakes indicates that the crust-fold becomes accentuated in form more than it advances along the surface of the fault which intersects it.

CHAPTER V

COMPLEX EARTHQUAKES AND THEIR ORIGIN

COMPLEX earthquakes are distinguished from simple and twin earthquakes by the long duration of the shock, the frequent changes of intensity and variations in apparent direction. They include among them the greatest earthquakes of which we have any knowledge, especially those which leave permanent traces of the movements to which they owe their origin. In these movements there is also considerable variety, and the three typical earthquakes considered in this and the two succeeding chapters are selected on account of the light which they throw on the different modes of origin of complex earthquakes. The Californian earthquake of 1906 originated in a single strike fault of immense length, the Mino-Owari earthquake of 1891 in two or more transverse faults crossing the main island of Japan, and the Assam earthquake of 1897 in a complicated system of faults probably connected with one great central thrust-plane.

The Californian earthquake of April 18, 1906, was in no way remarkable for great strength. The destruction wrought in San Francisco, serious as it proved to be, was due more to the fires that followed the earthquake and which spread unchecked owing to the rupture of the water-mains. The shock was evidently of considerable duration. Near the central region, the more sensible portion of the shock lasted a little more than a minute; but, according to one observer who was awake and at rest, the measured duration was not less than 3½ minutes. The strong vibrations, which formed the principal part of the shock, lasted about 40 seconds, and varied considerably in strength and direction. For instance, at Santa Rosa, the initial impulse was from the west, and, during the first portion of the earthquake, the motion was oscillatory, east and west. Then it became oscillatory, north and south, while finally there was a complex motion like that of a vessel in a choppy sea.

The area over which the shock was sensible extends from Coos Bay, Oregon, on the north, to Los Angeles on the south, the distance between these places being about 730 miles. To the east, it was felt as far as Winnemucca, Nevada, about 300 miles from the coast. The land-area disturbed was thus about 175,000 square miles, and the whole disturbed area, including the portion under the Pacific Ocean, probably about 373,000 square miles.

The area of maximum destruction is a band so narrow and of such length that it cannot be

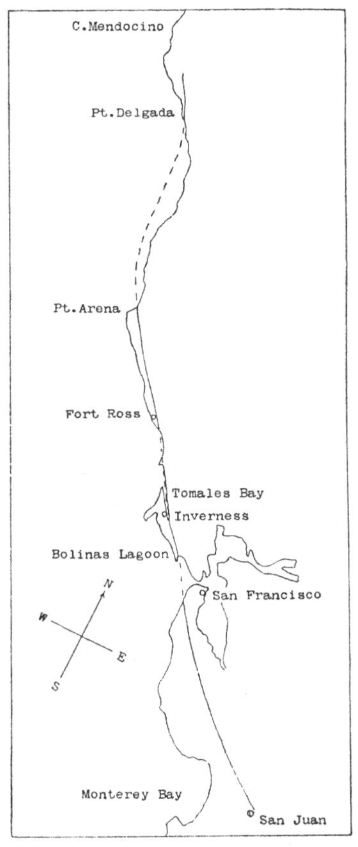

Fig. 10. Map of the San Andreas fault.

represented on a map on a scale small enough to be included in one of these pages. It clings closely (fig. 11) to the remarkable fault, the course of which is shewn by the continuous line in fig. 10, and indeed is almost exactly indicated by the width of the line there drawn. In no part does the width exceed two miles, and throughout its course the fault-line runs almost centrally between the two boundaries. The band extends from Cape Mendocino to east of Monterey Bay, a distance of 290 miles, except that, for about 70 miles from Point Delgada to Point Arena, both fault and band are interrupted by the sea; and the band maintains the same appearance almost throughout its entire length. When we consider the close relation between the band of maximum destruction and the course of the fault, it is impossible to doubt that the origin of the earthquake is to be sought for in this great fault.

That the fault was not a new one was at once evident. In parts, its true character had long been recognised; but it was not until after the earthquake that it was traced throughout almost its entire length. As at present known, it begins towards the north in Humboldt County, near Cape Mendocino, and it has been followed, with three interruptions, past San Francisco, to the north end of the Colorado Desert, or for a distance of more than 600 miles, and it is not impossible that it may reappear in another form still

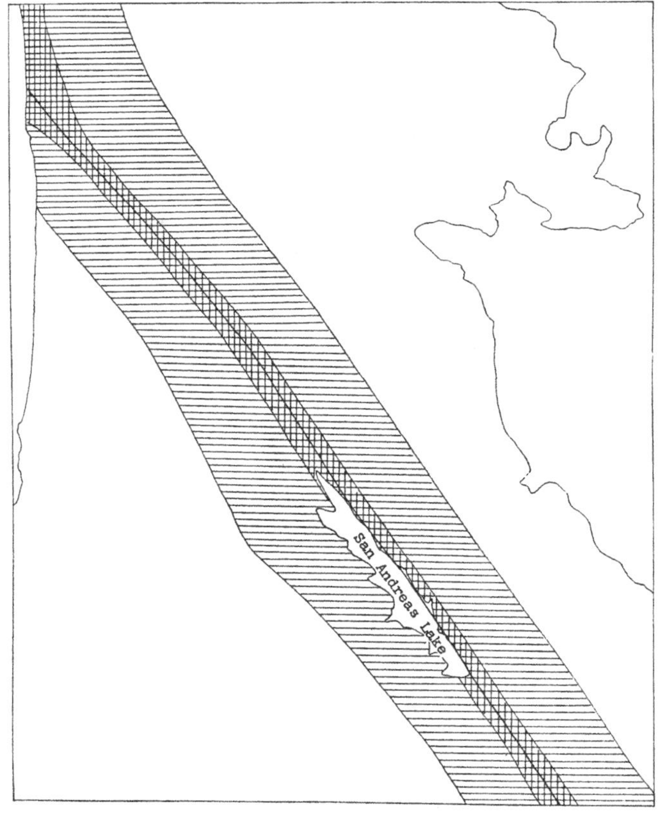

Fig. 11. Portion of the meizoseismal area of the Californian earthquake of 1906.

farther to the south. As will be seen from the sketch-map in fig. 10, it is never far distant from the coast. In three places it traverses the bed of the Pacific. From Fort Ross to Bodega Head and from Bolinas Lagoon to Mussel Rock, the breaks are short, being 13 and 19 miles respectively. Farther to the north, from Shelter Cove to Point Arena, the length of the sea-traverse is 72 miles; but, as will be seen presently, there can be little doubt as to the continuity of the fault throughout.

Next to its great length, the most remarkable feature of the fault is its approximate straightness, the average direction varying between 30° and 40° W. of N. When drawn on a map of small scale, such as that in fig. 10, the fault as a whole appears as a nearly even line, slightly curved and convex to the Pacific. On a large-scale map, however, it is seen that the fault is not a smooth uniform curve, but a succession of slightly curved rather than straight portions, the curvatures varying in direction.

Throughout its entire length, the fault lies along depressions or at the base of steep slopes, which are due partly to erosion and partly to displacement along the fault. Its position with regard to the mountain-ridge frequently varies, as it passes several times through breaks in the chains from the north-east to the southwest flank or *vice versâ*. Some remarkable depressions lie along its course, such as

Tomales Bay, a linear inlet more than 15 miles long and about a mile in width. The fault enters the San Francisco peninsula about eight miles south of the Golden Gate. In the first four miles, it is marked by a line of shallow longitudinal depressions, ponds and low scarps. Then the depression becomes more pronounced and passes into a steep, deeply trenched valley, 15 miles in length, the greater part of which has been converted by dams into the San Andreas and Crystal Spring Lakes for the water-supply of San Francisco. The name of the former lake has been given to the fault, which is known throughout its length as the San Andreas fault.

Evidences of recent movement are met with all along the fault. Scarps, or small cliffs, are common, formed by the elevation of one side or subsidence of the other or by both in conjunction. In some parts they are old, so worn down by the action of weathering, that only an experienced eye can detect them; in others they are still low steep walls, somewhat rounded, but yet bearing all the appearance of youth. In places, the scarps are double, including between them a trough-like depression. Small ponds are frequently met with, generally at the foot of scarps. Some have no outlet and the water in them is saline; others are mere pools formed on the line of pre-existing streams. In the rainy districts, these peculiar features of the growing fault tend to disappear more or less rapidly.

But in the southern portion, in the desert part of the Coast Ranges, they are more permanent. With every earthquake, the fissure opens anew; and, to the inhabitants of this district, the fault has become well known as the 'earthquake-crack.'

The movements which gave rise to the earthquake of 1906 were confined to the northern half of the San Andreas fault; but, with them, the characteristic features of the fault were renewed and intensified over a distance that is without a rival in our scientific records. Towards the south, the first traces of crust-displacement at the surface appear in the neighbourhood of San Juan, which lies 82 miles to the southeast of San Francisco. From this point, the fault-slip was followed without interruption, except for two short submarine portions, as far as Point Arena, a total distance of 190 miles. The fault then passes out to sea. But, farther to the north, in Humboldt County, the characteristic fault displacements reappear, and are seen for the last time in the neighbourhood of Cape Mendocino. Notwithstanding the length of the submarine portion to the north of Point Arena, and the discontinuity of direction at either end, the disposition of the isoseismal curves to the east of the submarine fault shows that the fault-line lies only a few miles from the shore and that it must run roughly parallel to the trend of the coast. There can therefore be little doubt that the fault observed near Cape

Mendocino is continuous with that which runs in a southeasterly direction from near Point Arena. The total length of the fault-slip at the surface thus cannot

Fig. 12. Fence severed by the San Andreas fault displacement.

fall short of 270 miles, or more than the distance from London to Newcastle and not less than that from London to Land's End. Nor is there any reason to suppose that even this amount represents the total length of the fault-slip below ground. The majority of earthquakes are caused by fault-slips of which no trace appears at the surface, and it is therefore possible that the displacement may have been continued underground for many miles at either end of the visible fault-movement.

For the most part, the fault-movement took place in a horizontal direction, though to the north of San Francisco some vertical displacement was also perceptible. There was a remarkable uniformity in the horizontal shift. It will be seen afterwards that the land on both sides moved, that on the southwest side of the fault to the northwest, and that on the northeast side to the southeast. The total horizontal displacement along the fault varied in different parts. As a rule, it lay between 8 and 15 feet, but, in places, the larger amount was exceeded, and at one place it was as much as 21 feet. This displacement was manifested at the surface by the dislocation of fences, roads, bridges, tunnels, pipes or any structure which crossed the line of the fault. An example of this shifting is given in fig. 12 which shows the shift of a fence near Bolinas Lagoon, and the flexure of the portion beyond the fault.

The appearance of the fault varies with the nature of the surface-rock. In hard ground, it is a simple crack or series of crevices; in loose soil, the ground is often bulged up so as to form a long narrow mound,

Fig. 13. New scarp (1906) of the San Andreas fault.

one or two feet high, and from five to ten feet wide. In many places, minor cracks branched off obliquely from the main fault for a few hundred feet or yards, and, in these, the zone of surface fracturing was several hundred feet in width. It is probably to these minor cracks, and to the resulting drag of the surface-beds, that we must attribute the variations in the amount of horizontal displacement along the main fault.

In all parts of the fault, the vertical displacement was small compared with that in a horizontal direction, and it was only distinctly manifested in the region to the north of San Francisco. The vertical displacement was shown principally in the formation of scarps, such as that near Inverness represented in fig. 13. Either new scarps were formed in places where none had existed before, or old scarps due to former movements were accentuated and renewed. Exact measurements of the height of the scarps were rendered difficult by the drag of the soil over the rupture, and this may account to some extent for the variations in height, which range from a few inches up to three feet.

Observations were made in those parts of the coast intersected by the fault, but they do not furnish conclusive evidence as to relative changes of level in land and sea. The Point Reyes Peninsula, on the southwest side of the fault, appears to have been slightly upraised, but the evidence does not amount to proof. On the other hand, the tide-gauge station in the Golden Gate lies on the northeast side of the fault, and observations made during the year after the earthquake show that the relative level of land and sea was the same after the earthquake as before. It would thus appear that the coast on the southwest side of the fault was up-raised during the great fault-slip.

From the appearance of the fault alone, it would be difficult to determine whether the crust on the southwest side was displaced to the northwest, or that on the northeast side to the southeast, or whether both sides moved at once. The point is one that could only be determined by a new trigonometrical survey of the district, and this was carried out during the year following the earthquake. The area covered by the survey, about 170 miles long and 50 miles wide at its widest point, extends from a line a short distance to the south of San Juan, where the fault-movements were first observed, to the neighbourhood of Fort Ross. When the calculations based on the new survey were completed, it was found that the various stations in the neighbourhood of the fault had all been displaced since the first survey was made, dating back in part to the year 1851. The estimated displacements varied from less than a foot to 19½ feet.

The re-survey is, in the first place, decisive in proving the important point that both sides of the fault were displaced, the southwest side to the northwest and the northeast side to the southeast. The displacements, again, were nearly always in directions parallel to the fault, and the amounts at stations close to the fault are comparable with those observed in lines of road or fencing severed by the fault. It is therefore evident that the movements revealed by the re-triangulation were in great part due to the sudden fault-slip which caused the earthquake of 1906.

Another point of great importance which the new survey has determined is the fact that the movements were not confined to the immediate neighbourhood of the fault. The crust to a distance of several miles on either side partook of the displacement though by amounts that diminished with increasing distance from the fault. For instance, on the east side of the fault, ten points at an average distance of a little less than a mile from the fault have an average displacement of 5·1 feet to the southeast; three points at an average distance of 2½ miles have moved on an average 2·8 feet in the same direction; while one point at a distance of four miles has a displacement of 1·9 feet. No point on this side at a greater distance than four miles has suffered any displacement distinctly exceeding that due to errors of observation, that is, none that can be detected with certainty. On the other hand, on the west side of the fault, twelve points at an average distance of 1¼ miles have an average displacement of 9·7 feet to the northwest; seven at an average distance of 3½ miles one of 7·8 feet; while one point (Farallon Lighthouse), 23 miles from the fault, was displaced 5·8 feet. Thus, in receding from the fault on either side, the displacement decreases more rapidly near the fault than at a greater distance; so that straight lines drawn on the surface on either side of the fault and at right angles to it would after the earthquake become slightly curved, the concavity facing the south on the northeast side of the fault, and facing the north on the southwest side. Moreover, for points on opposite sides of the fault and at equal distances from it, the displacements on the west side were twice as great as those on the east side up to a few miles from the fault, after which the ratio increases.

The relative displacement of the two sides of the fault, that is, the sum of the movements of the two sides, as detected by the re-triangulation, shews no variations in excess of what might be due to errors of observation, along the entire length of the fault from San Juan to Point Arena, except in a region near Colma to the south of San Francisco. If there were variations, they were less than could be detected by the means at the disposal of the surveyors. The earth-movements of 1906 were remarkable in many ways, in none perhaps more than in the regularity of their distribution.

The re-triangulation, as we have seen, assigns to the relative displacements along the fault a magnitude that is comparable with the observed shifts of the severed ends of roads and fences. But it is by no means certain that these shifts account for the whole of the estimated displacements. It is possible that some fraction of the total amount took place gradually and had for many years back been slowly increasing, though much the greater part must have occurred suddenly at the time of the earthquake. To produce the observed displacements in masses of so great a magnitude, it is clear that the seismic forces must have been for a long time gradually increasing in strength until they reached the point when they were sufficient to overcome all resistance. Before giving way, the crust may have begun to yield and bend in the directions in which it afterwards slipped. Then, quite suddenly, the slip took place over a large part of the fault and rapidly extended over the whole of the northern half of its course.

There is little need to seek for any other cause of the earthquake. In the immediate neighbourhood of the fault, the sudden spring of the displaced masses would inevitably shatter every building and fissure the surface soil. At a distance from the fault, as well as near it, the intense friction generated by such huge masses as they grated past one another would well suffice to produce an earthquake shock of the first magnitude. In the Californian earthquake of 1906, the origin of earthquakes was manifested in an unmistakeable manner. At the same time, the scale on which the operations of nature may be carried on was exhibited to a degree hitherto unknown in the annals of science.

CHAPTER VI

COMPLEX EARTHQUAKES AND THEIR ORIGIN (*cont.*)

THE superficial crust changes which took place in central Japan on October 28, 1891, are by no means the most remarkable known to us. The fault-scarp is exceeded in length by that formed during the Californian earthquake of 1906, in height by those of the Alaska earthquakes of 1899, and in complexity by those of the Assam earthquake of 1897. But the history of the fore-shocks and after-shocks and their connexion with the fault-scarp have been traced in greater detail than in any other earthquake. The minor shocks will be described in Chapters VIII and IX. The present chapter will be confined to an account of the fault-scarp and of the origin of the principal earthquake.

The provinces of Mino and Owari, in which the earthquake was chiefly felt, lie near the centre of the main island of Japan, and about 130 miles west of Tokyo. They are occupied for the most part by an extensive plain. Covered by a network of rivers and canals, this plain has been converted into one of the chief rice-producing districts of Japan, the largest towns in it being Nagoya, Gifu and Ogaki. The whole of it, however, is thickly populated, the road from Nagoya to Gifu, twenty miles in length, traversing a continuous succession of villages.

The meizoseismal, or most strongly shaken, area includes all this plain, as well as the mountainous district to the north, and contains about 4300 square miles. It is indicated by the shaded area in fig. 14. The northern half is a narrow band from four to eight miles wide. In the neighbourhood of Gifu it divides, a small branch proceeding to the southeast, while the main portion, twenty miles in width, continues past Nagoya to the south.

Within this area, nearly every structure raised by man was thrown to the ground or destroyed. The long street from Nagoya to Gifu was converted into a lane bordered by two long drawn-out lines of ruined houses. In the alluvial plain, especially in the neighbourhood of Nagoya, the ground was intersected by myriads of fissures. In some of the valleys depressions occurred, and houses suddenly sank up to their eaves.

From the hill-sides bounding the plain, great landslips fell, choking up the valleys, and long lines of mountains,

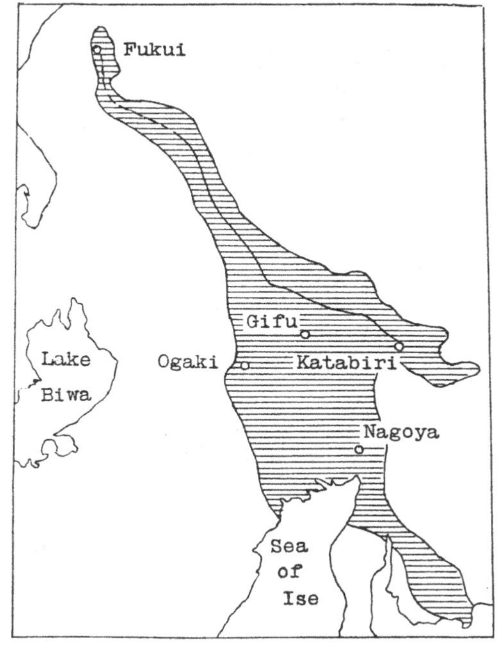

Fig. 14. Meizoseismal area of the Mino-Owari earthquake of 1891.

formerly green with forest, were laid bare. The railway-lines were generally disturbed, except where they passed through small cuttings. On the open plain, the ground between the sleepers rose in bolster-like ridges, as if it had been crushed. Evidences of compression were also manifested when the lines crossed small depressions in the plain or rose on embankments to the level of the bridges. The lines were then curved or crumpled. In the valley of the Neo, the ground is permanently compressed, and it would seem that the whole valley has become narrower. Plots of ground, formerly 48 feet in length, after the earthquake measured only 30 feet.

Earth-fissures, when they occur near rivers, are invariably parallel to the banks, and they are seldom of great length. They are evidently mere effects of the passing shock, and depend, for their magnitude and form, on accidental variations in the slope of the ground.

Very different from these local inconstant fissures was the great earth-rent of the Mino-Owari earthquake. It was distinguished from them in the first place by its length; it was actually traced continuously for a distance of forty miles, and, though only partially, for another thirty miles. It made its way independently of the slope of the ground, cutting across fields and hills indifferently, in a general northwest and southeast direction. But, more remarkable than the mere fracture was the distortion in its immediate neighbourhood. Almost uniformly throughout its whole length, displacements of two kinds occurred, one by which the crust on the northeast side was left permanently lower than that on the southwest side, the other by which the crust on the northeast side was shifted horizontally towards the northwest or that on the southwest side towards the southeast.

The course of the great fault is represented by the continuous line in fig. 14. At the southern end, it was first visible near the village of Katabiri, and from this point it was followed without interruption as far as the mountain Haku-san, that is, for a distance of forty miles. Still farther to the north, the fault crosses a range of mountains which were impassable during the winter months that followed the earthquake. At various places, however, a fracture possessing the characteristic features of the great fault, and no doubt continuous with it, was seen by other observers. The remainder of its course for the last fifteen miles at the northern end is somewhat uncertain, but the fault appears to end in the neighbourhood of Fukui, or about seventy miles from its starting-point at Katabiri.

The appearance of the fault varied considerably in different parts of its course. In some places there was no vertical movement, and the fault appeared as a mere crack in even ground, which might have escaped notice had it not been for the displacement of the northeastern side through a distance of three or four feet to the northwest. For instance, in a garden to the north of Gifu, two persimmon trees had stood for many years in an east and west line. The fault, however, passed between them from southeast to northwest, and, after the earthquake, the trees were found in a north and south line. In other places, roads and field boundaries were cut in two, and the severed ends displaced.

When the vertical displacement in the plains or valleys was small, the path of the fault was marked by a rounded ridge of soft earth from one to two feet in height (fig. 15), which closely resembled what might be produced by the burrowing of a gigantic mole. When the displacement reached a height of several feet, the fault-scarp formed a cliff which, being of earth, soon crumbled down into a slope. Nowhere was this effect more striking than at Midori in the Neo valley, about twelve miles northwest of Gifu. At this place the fault-scarp was twenty feet in height, and, from a distance, looked like a railway embankment (fig. 16). The northeast side, which was here the higher, was also shifted 13 feet to the northwest. By this latter movement, a new road leading to Gifu was cut in two obliquely, and the parts separated, as shown in fig. 16.

Throughout the whole length of the fault, whether the vertical displacement was considerable or imperceptible, the ground on the northeast side of the rent

Fig. 15. Fault-scarp, Mino-Owari earthquake of 1891. (By permission of the Council of the Royal Geographical Society.)

stands farther to the northwest than the ground on the opposite side with which it was formerly in contact. Generally, the displacement is about a yard or two, in some parts, as at Midori, it amounts to nearly 18 feet. The vertical displacement, with one exception, is also uniformly in one direction. The ground on the northeast side now lies lower than that on the other side, generally by a yard or two, in places by as much as 18 or 20 feet; the solitary exception to this rule is the striking one at Midori, illustrated in fig. 16, where the northeast side exceeds the other in height by 20 feet.

In addition to the great fault described above were many minor faults which apparently have not been mapped. The form of the meizoseismal area also points to the existence of at least one other great fault, of which no indication appears at the surface. The southern end of the fault-scarp lies in the portion of the meizoseismal area which branches off to the southeast. The main portion of that area continues to the southwest past Nagoya, and, though the destruction was less overwhelming there than in the neighbourhood of Gifu and in the Neo valley, it can hardly be doubted that the effects were due to a great fault-movement which died out before reaching the surface. Indeed, the greater width of the meizoseismal area would seem to indicate a greater depth of focus in this district.

Fig. 16. Fault-scarp at Midori, Mino-Owari earthquake of 1891. (By permission of the Council of the Royal Geographical Society.)

In Chapters VIII and IX we shall see how strongly this conclusion is supported by the evidence of the minor shocks. The earthquake was heralded during several years previously by many local shocks which originated in this area as well as in that bounding the future fault-scarp (fig. 19). It was followed by still more numerous shocks in the same district, and the evidence of those accompanied by sound shows that the depths of their foci gradually diminished on the whole with the lapse of time.

In the case of the Mino-Owari earthquake, we have no longer to deal with simple movements along a single fault. The displacement took place almost simultaneously along at least two great faults, which may have been connected in the neighbourhood of Gifu, but were probably different members of the same system of faults. The great fault-scarp was certainly not less than forty miles, and possibly seventy miles, in length. The evidence of the after-shocks shews that, towards the south, the fault-slip must have been continued for at least thirty miles at some depth below the surface, so that the total length of the displacement on that fault alone must have been about a hundred miles. In the main branch of the meizoseismal area, the subterranean fault-slipping must have extended more than fifty miles to the south of Nagoya to a point the distance of which from Fukui is about 110 miles. From the great length of the fault-slip, both at the surface and underground, it is thus clear that almost, if not quite, the whole width of the main island was involved in the movement.

We know too little of the actual conditions to form any detailed conception of the movements which gave rise to the Mino-Owari earthquake. From the seismic evidence alone, we are unable to determine which side moved in the neighbourhood of the great fault, whether the northeast side slipped to the northwest and generally downwards, or whether the southwest side was displaced to the southeast and generally upwards, or whether both sides moved in opposite directions or in the same direction but by varying amounts. No triangulation of the two provinces was made after the earthquake. Possibly there was no earlier survey of sufficient accuracy with which it could be compared.

The only evidence we possess on this subject comes from the neighbourhood of Midori. In this district, the river Neo runs on the east side of the fault. Above Midori, the river used to be a shallow rapid stream thirty yards wide and easily fordable. After the earthquake, it broadened out into a small lake, 70 yards wide, of still water which a boatman's pole could not fathom. This would seem to point to a real elevation of the rock on the northeast side, or, at any rate, of a much greater relative elevation near

Midori than farther up the river's course to the north.

There is other evidence of a general character, however, which throws some light on the problem. It will be seen in Chapter XII, dealing with the distribution of earthquakes, that, when a mountain-range such as the Himalayas, or a series of islands such as those which constitute the Japanese empire (fig. 26), is disposed in a curve or festoon, earthquakes are especially numerous and violent on the steeply-sloping convex side. The reason of this is that the crust-folds are being pressed forwards in this direction.

Now, the great fault or faults along which the displacements occurred in 1891 are transverse faults. They cut across the mountains obliquely, and roughly at right angles, to the crust-folds. If, as is probable, the movements which caused the earthquake were a continuation of those to which the present structure of the islands is due, then we may fairly infer that the whole crust in the neighbourhood of the faults concerned was thrust forward in a southeasterly direction.

Possibly the forces which so act upon the crust are uniform over a large area. But the masses into which the crust is cut up by the transverse faults offer varying resistances to movement, and consequently their displacements differ in amount. In the Mino-Owari district, the crust on the southwest side of the main fault advanced farthest; but, in the neighbourhood of Midori, owing probably to defective resistance, the crust on the northeast side was crumpled up and left in a position relatively higher than that on the southwest side.

It will be observed that a complex earthquake, like the Mino-Owari earthquake, differs in one important respect from the twin earthquakes described in Chapter IV. Both are due to differential movements along transverse faults, but twin earthquakes are due to the growth of a fold, complex earthquakes to the bodily displacement of the crust along a great length of the fault.

CHAPTER VII

COMPLEX EARTHQUAKES AND THEIR ORIGIN (*cont.*)

THE Assam earthquake of June 12, 1897, takes rank with the Lisbon earthquake of 1755 as one of the greatest of which we have any record. In disturbed area it has been equalled, if not surpassed, by others. It is in the vast extent of its seismic focus, in the complexity and amount of the distortion of the earth's crust within the central area, that it stands almost, if not quite, preeminent among modern earthquakes.

The district chiefly affected by this great earthquake lies some 250 miles to the northeast of Calcutta. The central area includes within it such places as Shillong, Tura, etc., but no large towns. To a great extent, it is covered by mountain-ranges,—by the Garo Hills on the west, and the Khasi and Jaintia Hills on the east,—and is uninhabited over wide areas. The full extent of the changes wrought in the earth's crust will probably remain unknown. But enough has been ascertained by the investigations made during the following winter to furnish some idea of their magnitude and complexity, and to reveal in broad lines the probable origin of the earthquake.

The sketch-map in fig. 17 shows the approximate form of the central area and the position of places within it where permanent changes in the crust were observed. The boundary, *A*, *A*, *A*, of the area should not be regarded as accurately drawn. For some parts of its course, the evidence is scanty or altogether wanting, but it is probable that it represents roughly the surface position and form of the region over which the movement responsible for the earthquake took place. There may, again, be many more places where distortions of the crust occurred. A large part of the district is covered with thick forests, which were impenetrable except by cutting paths through them, so that evidence could only be obtained from the neighbourhood of the beaten tracks.

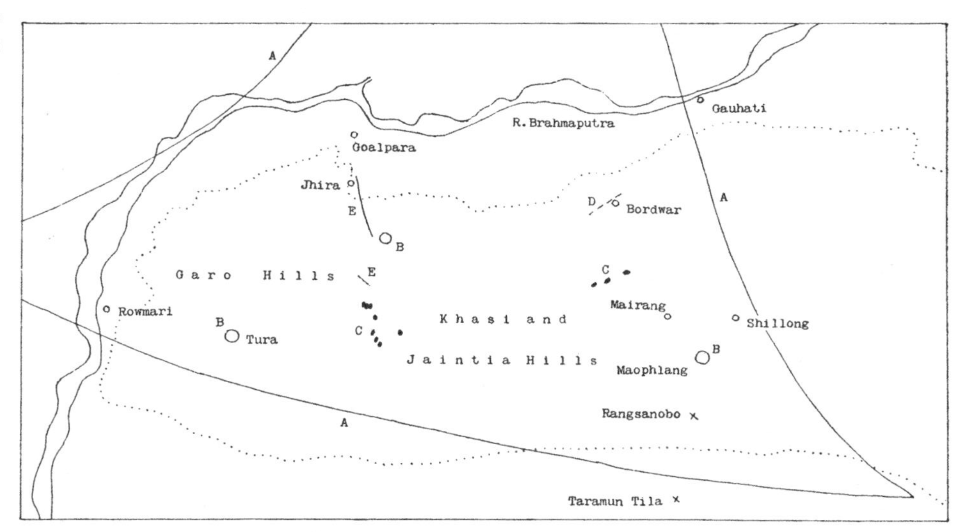

Fig. 17. Meizoseismal area of the Assam earthquake of 1897.

Assuming the boundary of the central region to be correctly drawn, its length must be about 200 miles and its greatest width not less than 50 miles. The area cannot be less than 6000 or 7000 square miles or about that of Wales, and may be as much as 10,000 or 12,000 square miles. To what depth the movement extended, we have no knowledge, but, even if it were not more than five miles, we can form some idea of the magnitude of the mass displaced, and of the forces required to produce so great a result.

The permanent changes which have taken place in the crust within this area belong to classes which pass gradually from one to the other. There are, first of all, mere bendings of the crust as shewn by visible changes in the form of the country and by differential variations of level without faulting. By these variations, the drainage of the surface is interrupted or diverted, and pools and small lakes are formed. In the second class are fractures, without any visible displacement of the surface-beds; and these pass by gradual transitions into the third class of faults.

Changes of the first and third classes, if they were to lead to distortions amounting to several feet in height, should obviously be measurable by a new trigonometrical survey of the district. If this could have been carried out over the whole area from a base-line of constancy beyond suspicion, the results would have been of considerable value and interest. The re-triangulation could only, however, be made over a limited district of the Khasi Hills, and, unfortunately, it is possible that the base-line may have undergone some change in length and position. The terminal stations of this line are Taramun Tila and Rangsonobo, indicated by small crosses in fig. 17. The former lies in the plain to the south of the central area, and its position and height may be regarded as practically unaltered. The latter is just within this area, and, consequently, its position may have varied. If it varied so as to result in lessening the distance between the two places, the effect of assuming the distance to be greater than it really is would be to increase the estimates of the lengths of the sides of all the triangles to the north. Now, this is the general result of the re-triangulation. The sides of every triangle, with one or two exceptions of little consequence, show an apparent increase in length greater than can be accounted for by errors of observation, and the apparent displacements of the stations, amounting to as much as 12 or 14 feet horizontally, increase as a rule from south to north, as they should do if the base-line were shortened. The heights of the stations also as a rule show an increase, but irregularly, the amounts in two cases being as much as 17 and 24 feet. Thus, at a first glance, the re-triangulation seems to point to an expansion and rise of the whole district covered by it. But the more probable explanation is that the base-line was shortened by a southerly movement of the station at Rangsonobo, and that the whole area under consideration was also displaced in the same direction. The variation in height was probably small, except at a few stations, and it is noteworthy that all of these lie close to fault-scarps or to places where movements are known to have occurred.

The results of the re-triangulation are thus inconclusive except as regards the fact that some change of level and position had taken place, and the probability that these were due to compression in a north and south direction. Somewhat more definite are some interesting observations indicating perceptible changes in the aspect of the hills (*B*, *B*, *B*, fig. 17).

The first of these is in the neighbourhood of Maophlang, in the Khasi Hills. Before the earthquake, the only part of the road to Mairang, another station 11 miles to the northwest, that could be seen was where it rounded a spur three miles distant; the road where it rounded the next spur being hidden by an intervening ridge. After the earthquake, a much longer stretch of the road round the nearer spur became visible, and it can also be seen rounding the spur beyond.

From a road over the Garo Hills, crossing the high ground near Cheran, it was formerly only just possible to see the river Brahmaputra over an intervening hill; after the earthquake, the whole width of the river became visible.

A somewhat similar observation was made at Tura, in the western portion of the Garo Hills. Signalling by means of the heliograph used to be carried on between the military police at Tura and Rowmari, a station 15 miles to the west on the east bank of the Brahmaputra. Before the earthquake, it was just possible to do this from a certain spot by means of a ray of light grazing an intervening ridge. After the earthquake, the difficulty vanished. Instead of Rowmari being just visible over the ridge, a broad stretch of the plains to the east of the Brahmaputra can be seen.

The changes in the aspect of the landscape, described in the last paragraphs, prove that local variations of level have taken place. They do not, however, admit of precise measurements. More definite are the surface undulations, unaccompanied by faulting, which have altered the gradient of drainage channels, and given rise to pools and small lakes. Two groups of these pools were observed, one in the Garo Hills, a short distance to the south of the fault-scarp to which reference will be made later, the other in the Khasi Hills, some miles to the south of the Bordwar fracture to be described later. They are represented by the black ovals, *C*, *C*, in fig. 17.

One of the group of pools among the Garo Hills may be regarded as typical of the class. It occurs in the valley of the Rongtham river, and is about a quarter of a mile in length. In outward appearance it does not differ from the pools which are common along the courses of mountain streams. Before the earthquake, however, no pool had existed in this spot, but merely a reach of the river strewn with rocks and boulders, exactly like that part of it immediately below the pool. In the direction that was formerly up-stream, the depth of the pool gradually increased, and its bed could be seen covered with coarse half-rounded boulders. Trees were still standing in the water, but killed by the recent submergence of the roots. About the middle of the pool, the greatest depth (of 12 feet) was reached, and just at this point the track from Darangiri crossed the stream, where formerly there was not more than a foot of water. Taking into account the original slope of the river-bed, the total relative change of level in the deepest part cannot be less than 24 feet. The formation of the pool was clearly due to an undulation in the surface-crust, for no trace of any faulting was perceptible.

Among the Khasi Hills, in the district covered by the trigonometrical survey, similar pools were formed, but on a smaller scale. One of them was about half a mile long and four feet in depth. The largest pool observed lay in the district between the Garo and Khasi Hills. This was more than a mile long, 150 to 200 yards wide, and its bottom could not be reached with a 20-foot pole.

In none of the districts containing these pools was there any evidence that their formation was due to faulting. About ten miles to the north of the pools among the Khasi Hills, however, the undulations of the crust give place to fracturing. The principal fracture is one near Bordwar, indicated by the broken line *D* in fig. 17. It can be traced for a distance of seven miles, being clearly visible on the sides and summits of the hills which it crosses, and leaving unmistakeable traces of its passage over the intervening low ground and forests. The fracture itself is only a few inches wide. It is distinguished from the fault described later by the absence of any displacement, either vertical or horizontal. Nevertheless, like the faults, the Bordwar fracture is clearly connected with the origin of the earthquake.

In the immediate neighbourhood of the fracture, the violence of the shock was extreme. Trees were overthrown or snapped across, great slabs of rock were rent in two, and large boulders were dislodged from near the crests of the hills and scored pathways through the forests with which the hillsides were clothed.

The fractures without displacement pass by

insensible gradations into the fractures with displacements or faults. The number of small fractures and faults is unknown. Of faults of considerable size, however, there appear to be only two, which are indicated by the straight continuous lines *E*, *E*, in fig. 17, a small one in the neighbourhood of Sámin, the larger along the course of the Chedrang river. The Sámin fault is about 2½ miles in length, with a maximum throw of 10 feet. As will be seen from the sketch-map in fig. 17, it lies about three or four miles to the north of the group of pools in the Garo Hills.

The Chedrang fault is of greater importance. Its southern end is close to the spot mentioned on a former page (p. 84) as that from which changes in the aspect of the Brahmaputra were perceptible. It has been traced for a total length of 12 miles, but, in both directions, its termination is uncertain; at the north end it is lost beneath the alluvium of the Brahmaputra, at the other in thick jungle. It is represented on a larger scale by the broken line in the accompanying map (fig. 18). In this, the singular straightness of its course is manifest. The figures on the right-hand side indicate the throw in feet of the fault in different parts of its course. In two places, there is no vertical displacement; in another, it amounts to as much as 35 feet. But, wherever any displacement is perceptible, it is invariably the east

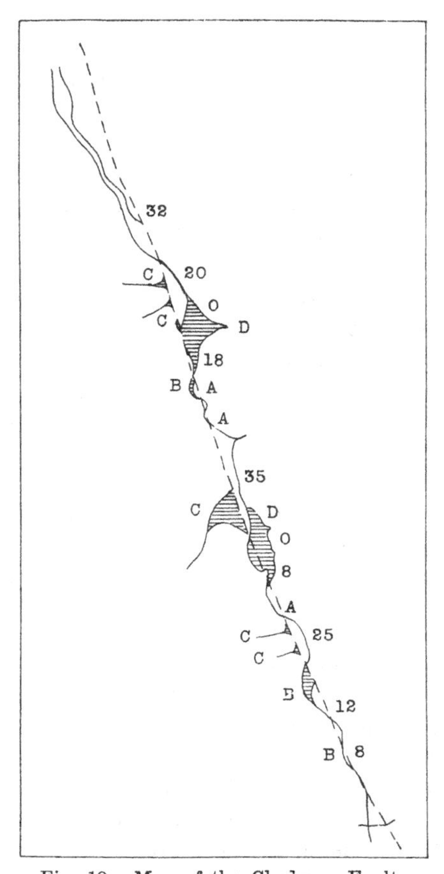

Fig. 18. Map of the Chedrang Fault.

side which is higher than the other. Two other points may be noticed here. The fault differs in one important respect from those in California and Japan. The throw, great as it is, is unaccompanied by any horizontal displacement, the movement being entirely vertical. Again, wherever the plane of the fault could be seen in rock, it was practically vertical. There was no decided inclination in one direction or the other.

The general direction of the fault is from south-south-east to north-north-west, and, throughout its known course, it runs along the valley of the Chedrang river. The river flows from south to north. As the rock, when there is any displacement, is always higher on the east side, it follows that, whenever the stream crosses the fault-line from east to west, there is a waterfall, as at *A*, etc. When the stream meets the fault from the opposite direction, it is ponded back and forms pools, as at *B*, etc. Pools also collect on the west side of the fault when small tributary streams meet the scarp, as at *C*, etc.

The largest pools, however, are those at *D*, *D*, the more southerly of which is half a mile long and 300 or 400 yards wide. The pool spreads across the fault, and there is no barrier visible. The channel of the stream gradually sinks in the upstream direction, without abrupt change, beneath the waters of the pool. Now, it is just where the fault has no throw that the pool is widest and deepest. The pool is therefore due to the formation of an undulation in the surface of the ground, by which the natural slope of the channel has been reversed. A similar pool occurs a little more than a mile farther down the stream. Here, again, the throw of the fault vanishes; but, the valley being more open and the gradient lower, the pool is somewhat larger, being half a mile in length and nearly the same in width. Here, too, the stream and lake merge into one another gradually; there is no faulting or abrupt change of level, merely an undulation like those which resulted in the formation of the pools among the Garo and Khasi Hills.

Not far from its northern end the fault passes beneath a thick bed of alluvium. The fault-scarp disappears and is represented by a steep slope of the surface. When the fault enters the open plain, this slope in turn disappears, but the effect of faulting is still noticeable in the formation of a large pool near Jhira, a mile and a half long, three-quarters of a mile wide, and about 12 feet deep. The recent formation of the lake is evident from the large number of trees still standing which have been killed by the submergence of their roots.

Along the course of the fault-scarp, there is no evidence to show whether the rock on the east side had risen or that on the west side had subsided. The barrier which gave rise to the lake near Jhira must, however, be due to the elevation of an undulation in the alluvium, for the stream issuing from the lake now runs down a steep bed of alluvium. The whole of the alluvial plain of Lower Assam is raised so little above the level of the sea and the gradient of the rivers is consequently so small, that any depression sufficient to account for the steep slope of the issuing stream would have resulted in extensive flooding of the plains. It therefore seems probable that the fault-scarp is due mainly to the elevation of the east side of the fault rather than to the depression of the other side.

It will be noticed, on referring to fig. 17, that the places where the different changes described above occurred are scattered over the central area, not altogether irregularly, but with some approach to order. In the south, the changes are probably exhibited as long low rolls. Farther north, these undulations become more pronounced. They are sufficient to reverse the drainage of even rapidly-flowing streams. Still farther northwards, the undulations are replaced by fractures and faults. Beyond the Garo and Khasi Hills the crustal changes decline in magnitude and abruptness. But they probably extend beneath the bed of the Brahmaputra, for it is difficult to account otherwise for the floods and changes of level which have taken place along this valley.

The changes referred to are only those of the surface. They must, for some distance, increase in magnitude with the depth. Fractures, which at the surface show no displacement, may be faults a mile or two below it. The undulations which give rise to pools along stream-courses in all probability are mere folds in the upper rocks produced by faulting below. It is only when the displacement occurs at a small depth that fault-scarps are left projecting at the surface.

The remarkable feature of the Assam earthquake is thus the large number of apparently independent centres of disturbance. That they were to a great extent independent, once the earthquake was over, is clear from the fact that they became more or less isolated centres of after-shocks. But it is equally clear that, over the whole central area, the movements which caused this great complex earthquake took place nearly, if not quite, simultaneously. The faults and incipient faults scattered over this area must therefore be connected as off-shoots or branches with a fracture occupying the greater part of the central area, and yet in no place at any considerable depth (possibly not more than five miles or so) below the surface.

The only fracture of this kind with which we are acquainted is that known as a thrust-plane. Beneath an area not less than that of the whole of Wales, a great thrust-plane seems to extend, with the off-shoots or branch-faults reaching up to the surface. The earthquake seems to have been mainly due to a forward bodily slide of this immense mass of rock towards the south, crumpling up or fracturing the surface-layers, and, by numerous displacements along the minor branching faults, forming local centres of violent action. Such a crust-slide, it is evident, must have produced an earthquake, in nature most complex, of a strength sufficient to overthow the greatest works of human hands as easily as the frailest.

CHAPTER VIII

FORE-SHOCKS AND THEIR ORIGIN

OCCASIONALLY a great earthquake occurs without the slightest warning of its coming. Before the Californian earthquake of 1906 and the Messina earthquake of 1908, no slight shocks appear to have been observed, at any rate in the central districts. Frequently, however, some notice is given in the form of feeble shocks felt here and there in the central area or of tremors that render birds and animals restless and uneasy but remain imperceptible to man. The great Charleston earthquake of August 31, 1886, was preceded by slight but distinct shocks on August 27, 28 and 29, possibly by others in the previous month of June. Six slight shocks occurred within ten hours before the Riviera earthquake of February 23, 1887. Besides these undoubted shocks, faint tremors seem to have been perceived over a wide area by animals and birds.

In the earthquakes of this country there is the same variety. Some are preceded by slight shocks, others by none. The Inverness earthquake of 1901 was heralded by a slight shock. This was also the case with its predecessor in 1890, and with the strong earthquakes of Pembroke in 1892, Hereford in 1896, Carnarvon in 1903 and Derby in 1904. On the other hand, the Pembroke earthquake of 1893, the Derby earthquake of 1903 and the Swansea earthquake of 1906, all occurred without the slightest warning being given.

In some, possibly in many, cases the absence of warning may be more apparent than real. A study of the preceding conditions may show a marked increase in frequency of slight preliminary movements. The Mino-Owari earthquake of 1891, for instance, seemed to have come unannounced, except for a rather strong shock which occurred three days before. But in reality the preparation for the earthquake had been going on for several years. In the provinces traversed by the great fault-scarp described in Chapter VI, slight shocks were not infrequent.

From 1885 to 1889, the number which occurred in the immediate neighbourhood of the fault was five times as great as in an equal area farther away. During the years 1890 and 1891, up to the day of the earthquake (October 28), the average frequency of earthquakes along the scarp became ten times as great as elsewhere.

Besides their increase in frequency, the distribution of these earlier shocks marked out the future lines of disturbance. When the centres of all the shocks during the years 1890 and 1891 (up to October 28) are plotted on a map, curves may be drawn through the centres of equal areas which contain the same number of earthquake centres. The dotted lines in fig. 19 represent the coast-line, the broken line the course of the great fault-scarp described in Chapter VI, the continuous lines are the curves referred to. The meaning of any one of these curves, say that marked 5, is that if any point on the curve be regarded as the centre of a rectangle, the sides of which are one-sixth of a degree of latitude and longitude in length, then five centres are situated within this rectangle. It will be seen that the curves follow the line of the great fault and its continuation, while, from about the centre of the map, there branches off another group of curves towards the south. We may infer from these latter curves that there was at least one other fault in action besides that along which the

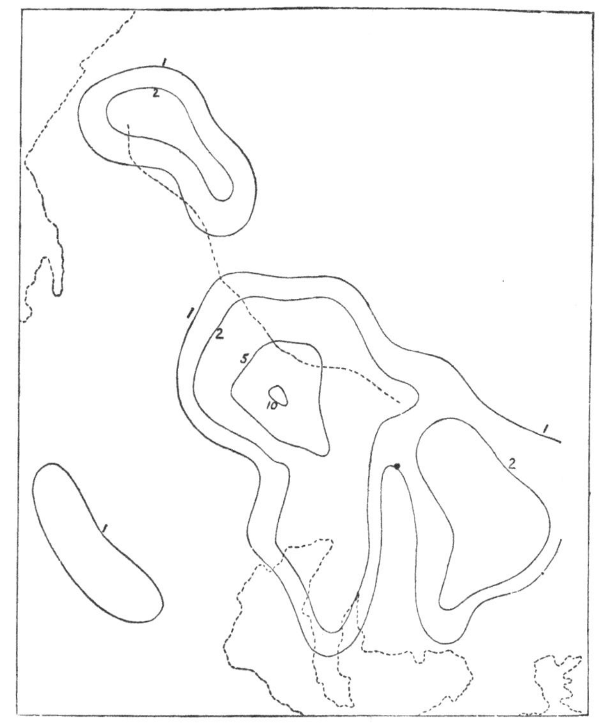

Fig. 19. Distribution of the Fore-Shocks of the Mino-Owari earthquake of 1891. (By permission of the Council of the Geological Society.)

great scarp was formed—a conclusion that will be found confirmed in the next two chapters.

Thus, along the line of the fault, for at least two years before the earthquake, shocks had been taking place, not only with greater frequency, but with closer concentration along the region of the future earthquake. Had these preliminary symptoms been recognised, they might have served as some, though perhaps a very uncertain, warning of the great catastrophe.

Whether they be few or many in number, the origin of these fore-shocks is probably the same. The strain, which ultimately results in a great fault-slip, is the growth of many years, possibly of centuries. The resistance offered to the movement is not uniform throughout the fault-surface. Here and there are obstacles which must first be overcome. Small slips take place in these regions, each giving rise to a fore-shock. Their chief purpose is to equalise the effective resistance to motion over the fault-surface, so that, when at last the strain increases and the climax comes, the movement takes place instantaneously or nearly so throughout a great extent of the fault.

Sometimes, however, the obstacles appear to be absent, and the great movement takes place without any apparent minor slips. But a more careful study, a greater knowledge of the details, would probably reveal that some preparation had been taking place.

CHAPTER IX

AFTER-SHOCKS AND THEIR ORIGIN

THE parts which after-shocks play has been briefly touched on in the chapters which deal with simple and twin earthquakes. In the cases considered, the after-shocks were few in number, and they could be followed without difficulty. They were found to be due to the additional strains brought into action by the movement which caused the principal earthquake. They testified to the gradual relief of this strain, especially along the lateral and upper margins of the principal focus, until the displaced rock-mass once more attained equilibrium throughout its whole extent.

In the great earthquakes, which have formed the subjects of Chapters V–VII, after-shocks were far more numerous. In the central area they were usually so frequent as to defy all attempts to chronicle them. For many hours, even for several days, the ground seemed never to be at rest. So incessant were the tremors that the surface of the water in a glass maintained a constant ripple, lamps and chandeliers did not cease to swing. Many after-shocks attain a fair degree of strength. They may help to complete the destruction left unfinished by the parent-shock. Rarely, however, do the fresh shocks exceed the first in severity. They show a rapid though fluctuating decline, both in frequency and intensity, until, after the lapse of a few years, the normal frequency of earthquakes in the district is once more attained.

As a rule, the same laws govern the distribution in time and space of all great earthquakes, at least so far as they are known. The decline in frequency of after-shocks has been studied in many earthquakes, their distribution in space has been investigated in only a few cases, the materials for the purpose being generally insufficient. In no country are the observations so abundant and exact as in the empire of Japan, and of no earthquake in that country have the after-shocks been studied with such care as in the great Mino-Owari earthquake of October 28, 1891.

This great earthquake, which is described in Chapter VI, occurred at 6.37 a.m. (mean time of central Japan). The course of the remarkable fault-scarp is shown in fig. 14, Gifu, the capital of the province of Mino, being only about 4½ miles from the fault-scarp. The seismograph at the meteorological observatory in this city recorded the first half-dozen vibrations, and then the building fell and the instrument was buried though uninjured. At 2 p.m., or little more than seven hours after the earthquake, it was again in working order, and the registration of the numerous after-shocks began. From this time until the end of December 1893, that is, in about two

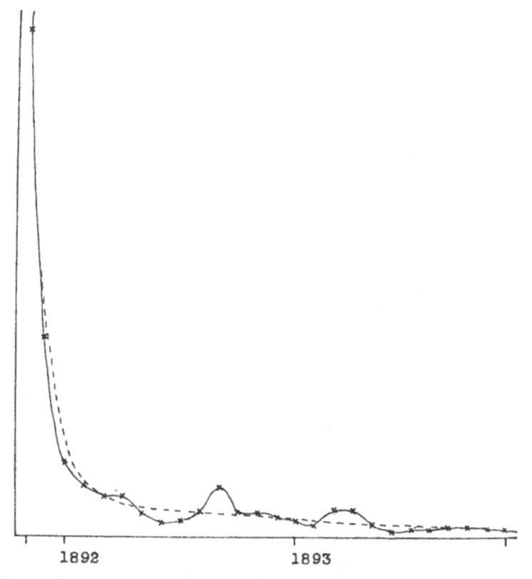

Fig. 20. Frequency of the after-shocks of the Mino-Owari earthquake of 1891 at Gifu.

years and two months, the total number of after-shocks recorded is 3365. Of these, 10 are described as violent, 97 strong, 1808 weak and 1041 feeble, while the remaining 409 were merely earth-sounds unaccompanied by any perceptible tremor.

The relation between the number of after-shocks and the time since the great earthquake is shown graphically in fig. 20. The numbers of after-shocks in successive months from November 1891 to December 1893 are represented by the distances of the dots from the horizontal line. The continuous curve, which passes through these dots, shows how rapidly the after-shocks declined in frequency during the first few months. Indeed, in November 1891 there were 1087 after-shocks, and in the following month 416. A year later, in December 1892, the number had fallen to 39, and two years later, in December 1893, to 16.

It will be noticed that the decline in frequency is not continuous. Violent shocks occurred, for instance, on January 3 and September 7, 1892, and each of these was followed by its own train of after-shocks, thus causing temporary fluctuations in the total number. Smoothing away the irregularities in the curve, we get the dotted curve in fig. 20, which in all probability represents the true law of decline in frequency. Assuming this to be the case, it has been estimated that the ground will not recover its normal stability, that is, the annual number of shocks at Gifu will not have returned to the previous number of 15, until after the lapse of nearly forty years. By this time, the total number of after-shocks recorded at Gifu will be very nearly 4000.

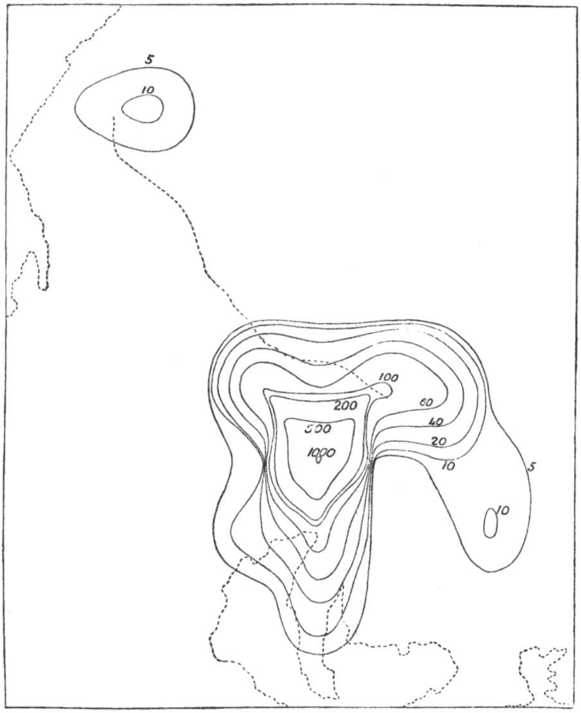

Fig. 21. Distribution of the after-shocks of the Mino-Owari earthquake of 1891, Nov.-Dec. 1891. (By permission of the Council of the Geological Society.)

In addition to the decline in frequency represented in fig. 20, there was at the same time a rapid but fluctuating decrease in strength. Of the ten violent shocks recorded at Gifu, nine occurred within the first four months, and the last in September 1892. All of the 97 strong shocks occurred within the first thirteen months, all the weak ones but four within the first twenty months. During the few days immediately after the earthquake, feeble shocks and sounds were rarely noticed, and it was only after the lapse of several weeks that they came into prominence. Towards the close of 1893, with the exception of four weak shocks, feeble shocks and sounds were alone recorded.

The Japanese records are so detailed that it is possible to trace not only the decline in frequency of the after-shocks, but also the migrations of their foci along the great fault. To represent graphically the movements of the individual foci, or even to try and follow them mentally, would only lead to confusion. The number of earthquakes is too great. To discover the general laws of distribution of the foci along the fault, the after-shocks must be grouped in periods of some duration.

In the next two diagrams (figs. 21 and 22) an attempt is made to represent the distribution of the after-shocks in two-monthly intervals, fig. 21 corresponding to the months of November and December

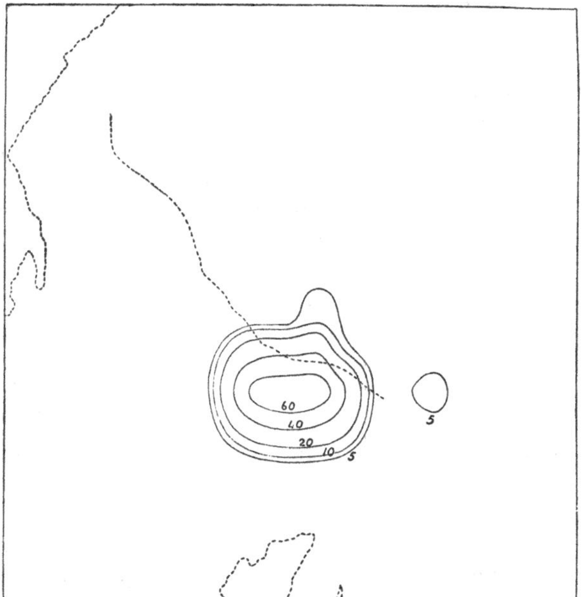

Fig. 22. Distribution of the after-shocks of the Mino-Owari earthquake of 1891, July-Aug. 1892. (By permission of the Council of the Geological Society.)

1891, immediately following the great earthquake, fig. 22 to the months of July and August 1892. The maps are constructed in the same way as that in fig. 19, and the lines have the same meaning.

The two principal features of these maps are evident at a glance. The first is the much greater number of earthquakes during the earlier interval; the second, and more important from our present point of view, is the much wider distribution of the earthquake centres in this interval.

From fig. 21 we learn that in November and December 1891 after-shocks occurred at the northern end of the fault-scarp, though principally in its central and southern regions. They shew also that slipping took place along the continuation of the fault-scarp towards the south, though chiefly along another fault of the existence of which we have already seen evidence in the preceding chapter. Fig. 22 shews the shrinking of seismic activity towards the central region with the lapse of time.

The distribution of after-shock centres thus indicates the existence of a nearly central region of extraordinary activity, and of two more or less isolated districts near or surrounding the extremities of the fault.

The seismic activity of these terminal districts was not only less marked, but was also of shorter duration, than that of the central district. At the northern end of the main fault, as well as at the southeastern end of its continuation, all action practically died out before April 1892. In the southern terminal region of the fault-scarp it lasted until, if not after, the close of the same year. A similar withdrawal of action from its southern extremity characterises the secondary fault, only two centres lying in its neighbourhood after March 1892.

The after-shocks of the Mino-Owari earthquake thus followed the same laws as were afterwards exhibited on a smaller scale by the Inverness earthquake of 1901, namely, a general decline in frequency, decrease in the area of seismic action, and a gradual but oscillating withdrawal to a more or less central district.

Whether the numerous after-shocks of the Mino-Owari earthquake were or were not accompanied by any change in the height of the fault-scarp is unknown, for measurements of its height at different times do not appear to have been made. Some light, however, is thrown on this point by the great Concepcion earthquake of 1835. This earthquake was probably caused by a displacement along a submarine fault running in a direction parallel to the neighbouring coast-mountains. That it was accompanied by the formation of a fault-scarp is implied by the occurrence of the sea-wave which afterwards swept the adjoining shores. The coast was also raised by several feet, at one point by not less than ten feet. The earthquake was followed by hundreds of after-shocks, some of considerable violence, proceeding apparently from the same origin; and, concurrently, the coast subsided, for, after an interval of some weeks, it stood at a lower level than it did immediately after the principal shock.

The distribution of the after-shocks of the Mino-Owari earthquake was simple compared with that of the shocks which followed the Assam earthquake of 1897. The central area in this case, as has been seen, covered a tract of at least 6000 or 7000 square miles. Within it were two fault-scarps, one fracture without perceptible displacement, many smaller faults and fractures, and numerous undulations of the surface-crust. At Shillong, Tura and other places within this area, the after-shocks during the day following the earthquake were to be numbered by hundreds. Indeed, for days afterwards the ground seems to have been in a continual state of tremor. Interspersed among them were several strong shocks that would have caused destruction in the central area if there had been anything left to destroy. Three of these on June 13, that is, the day after the earthquake, were felt in Calcutta, 250 miles from the centre; one on June 14 was felt in the same city, and three others on June 29, August 2 and October 9.

In mere frequency, the after-shocks of the Assam earthquake must have far surpassed those of the Mino-Owari earthquake. At Shillong and Tura they were so numerous as to defy all estimate of their number. Up to the end of June 15, that is, in little more than three days, 83 were recorded at Maimansingh and 561 at N. Gauhati. During the following month (June 16–July 15), 98 were felt at Maimansingh and 209 at N. Gauhati. Non-instrumental records are inevitably incomplete, so that the real numbers must have been far greater. Again, from August 1 to 15, 124 shocks were felt at Tura, 151 at Darangiri, and 182 at Goalpara. At Maophlang, 1050 shocks were counted from October 1, 1897, to September 30, 1898, and at Mairang 841 in the same interval.

More interesting, from the point of view of their origin, is the fact that, at all of these places, the majority of the shocks were very slight, and consequently of local origin. Maophlang and Mairang, for instance (fig. 17), are only eleven miles apart. At the former place, 92 after-shocks were felt from September 12–28, and at the latter 83 in the same interval. But, of the after-shocks at Maophlang, 37 are described as smart, 45 slight and 10 feeble; of those at Mairang, 6 were regarded as smart, 10 slight and 67 feeble. Owing to the difficulty of obtaining correct time, the identity of the shocks at both places cannot always be established. Regarding shocks as identical, however, if their recorded times of occurrence do not differ by more than 15 minutes, there were in the interval considered 19 common to both places, leaving 73 as peculiar to Maophlang and 64 to Mairang. Moreover, of the 19 common shocks, only one is described as smart at both places, three were smart at one and slight at the other, and nine were smart at one and feeble at the other. The remaining six shocks were slight at one place and feeble at the other, and therefore the records probably do not correspond. The conclusion to which these figures lead is that the great majority of the after-shocks recorded at any place are of local origin, and that, all over the vast central area of the Assam earthquake, there were scattered centres of activity, which were apparently independent of one another, although some law, of which we have no knowledge, may have governed the relative frequency in the different centres.

An interesting observation was made at Maophlang by the recorder of the after-shocks. A straight piece of wood was nailed to a stout post so that its upper edge pointed exactly to the crest of a ridge about 1½ miles to the west. Six months later, this edge pointed some way down the slope of the ridge. The apparent angle through which it had been tilted was about one degree. The change might be due to a displacement of the post, of which, however, there was no evidence. It probably implies that crustal displacements continued long after the great earthquake, and that they were, in part at any rate, due to the slips which caused the stronger after-shocks.

The general result to which these observations of after-shocks leads is that the displacement of the earth's crust which gives rise to a great earthquake at once changes the conditions of strains to which that and the surrounding portions of the crust are subjected. The sudden increase of strain is relieved by small slips here and there along the fault-surface. If the initial displacement be confined to one or two faults of a system, the area of this displacement at first extends outwards; then it leaves the terminal regions and shrinks continually towards the central region, to which it becomes confined at the close of the operations. In a complex earthquake, like that of Assam, the conditions are extremely complicated. In this case similar changes are introduced not only in the parent-fault but in all the minor branch faults, and thus innumerable after-shocks occur in many widely-separated portions of the central area.

From the mode of their formation, it is evident that after-shocks must be far less violent than the parent-shock. This is almost invariably the case, but there are occasionally exceptions when the strain is not completely relieved by the first great movement. The Zante earthquake of January 31, 1893, was followed on April 17 by a still stronger earthquake; and this was also the case in northeast Greece in 1894, when the earthquake of April 20 was succeeded by the more violent earthquake of April 27, which was accompanied or caused by the formation of a fracture 34 miles in length.

CHAPTER X

SYMPATHETIC EARTHQUAKES AND THEIR ORIGIN

THE origin of after-shocks has been traced in the last chapter to the increased strains along a fault which are brought into action by the great movement in which the principal earthquake originated. But these strains are not confined to the region of the fault alone. In the whole surrounding country, the strains to which the crust is subjected must be different after a great earthquake from what they were before. They may be increased or lessened by the displacement which gave rise to it, and the result may be either an increase or diminution in the seismic activity of neighbouring regions.

For instance, during the months which followed the Inverness earthquake of 1901, there were 15 undoubted after-shocks which originated in the same fault as the principal earthquake. In the valley of the Findhorn, which lies 13 or 14 miles to the southeast of the Great Glen fault, at least two earthquakes, and possibly others, were felt during that interval, which were not perceived by any of the observers in the neighbourhood of Inverness. It is probable, therefore, that they were local shocks, and, as earthquakes peculiar to the Findhorn valley are rare or unknown, it would seem that the slips which gave rise to them were precipitated by the movement or movements along the Great Glen fault—in other words, that they were sympathetic shocks of the Inverness earthquake.

Clearer evidence is furnished by the districts which surround that in which the Mino-Owari earthquake of 1891 originated. The central district is shown in fig. 14. About 45 miles to the east, and 55 miles to the west, of the great fault-scarp are two other districts in which earthquakes are somewhat frequent. In the eastern district 29 earthquakes, and in the western district 20 earthquakes, originated between January 1, 1885, and October 27, 1891. After the earthquake, from October 28, 1891, until the end of 1892, the numbers which originated in the same districts were 30 and 36, or, in November 1891 alone, 7 and 8. Thus, for every earthquake in the eastern district before October 1891, 6 were felt in the interval afterwards and 10 in the month of November alone; for every earthquake in the western district before October 1891, 10 were felt in the interval afterwards, and 16 in November alone.

In this case, however, the increase of seismic

activity need not be a consequence of the great displacement. The shocks in both central and adjacent districts, it is possible, might result from a general increase of strain over a wide extent of country, and the augmented frequency in the lateral districts could not with justice be regarded as an undoubted effect of the former, for they might both be effects of the same cause. But that the connexion is one of real dependence is probable for two reasons. Crustal distortions of the kind and magnitude which took place in the Neo valley could not be effected without a very considerable change of strain in all the surrounding country. An increase of strain, again, cannot determine the occurrence of an earthquake unless it be sufficient to overcome the resistance to displacement. Now, it is unlikely that the gradual increase of strain should be so nearly proportioned everywhere to the prevailing conditions of resistance as to give rise to a marked and practically simultaneous change in seismic activity over a large area; whereas the sudden occurrence of a strong earthquake might alter the surrounding conditions with comparative rapidity, and induce a state of seismic excitement in the neighbourhood. The rapid and simultaneous increase in earthquake-frequency in the two subsidiary districts, distant though they be from one another by a hundred miles, seems strongly in favour of this interpretation.

CHAPTER XI

EARTHQUAKE-SOUNDS AND THEIR ORIGIN

THE sound which accompanies an earthquake is a deep rumbling noise, often described as a dull heavy boom, a heavy groaning noise, a low moaning, or a low heavy rumble. As a rule, it is compared to some well-known or definite type of sound, such as waggons or traction-engines passing, a peal of thunder, the rising of the wind, loads of stones falling, the thud of a heavy weight, an explosion, or to miscellaneous sounds such as the trampling of many animals or the rumbling of the sea in a cave, etc.

The most important feature of the earthquake-sound is its extraordinary depth, as if it were too low to be heard, and, with many persons, this certainly is the case. Much lower than the lowest thunder is the description of one observer; of others, that it was a rumble that could be felt, or a noise more felt than heard. The same impression is conveyed by the frequent use of the word 'heavy,' in such comparisons as a very heavy steam roller passing, heavy peals of thunder, heavy gusts of wind, a heavy fall of rocks in a quarry, a heavy blast, or the heavy rumbling of sea-waves in a cave. On an average, one out of every three observers qualifies his description by the use of the word 'heavy.'

Still more decisive is the evidence furnished by the fact that the same sound is heard by some observers and not by others. One will describe the sound as like the rumbling of a heavy traction-engine passing or as louder than any thunder; another in the same place will be equally positive that the shock was unaccompanied by sound. Not only people in the same town, but persons in the same house, and even in the same room, differ in this respect.

A popular explanation of this partial inaudibility of the sound is that it is due to inattention. But the sound is not heard sometimes by those whose hearing is keen for ordinary sounds and who are listening intently at the time. Moreover, in the Inverness earthquake of 1901, which occurred at 1.24 a.m., the sound which preceded the shock was heard by 86 per cent. of the observers who were awake, and by 84 per cent. of those who were asleep, when the earthquake began. In the Doncaster earthquake of 1905, which occurred at 1.37 a.m., the corresponding figures are 93 and 91 per cent.

In the neighbourhood of the centre, the sound varies greatly in character and intensity. It grows rapidly louder as the preliminary tremors increase in strength. When the principal vibrations begin, the sound becomes deeper and more rumbling, and with the strongest vibrations deep booming explosive crashes are sometimes heard. The exceeding depth of the sound is manifested in the variability of the impression which it produces. At one place, the sound, according to one observer, will begin like a rushing wind and culminate in a loud explosive report; another will hear a noise like distant thunder which ends as the shock begins; while a third will hear no sound at all. In the zone outside the central isoseismal line, the crashes are seldom heard, but still the sound changes perceptibly when the shock is felt, becoming rough and grating; while near the boundary of the sound-area it is a low monotonous moan like the boom of very distant thunder.

Occasionally, the sound is heard before the shock and becomes inaudible as soon as the first vibrations are felt, or begins as the shock ceases to be sensible and afterwards dies away. As a general rule, however, the sound accompanies the shock, and is heard before it by a large number of persons, and after it by a somewhat smaller number. Taking British earthquakes only, 64 per cent. of the observers hear the sound which precedes the shock, while 43 per cent. hear the sound which follows it. For all earthquakes, whether strong, moderate or slight, these proportions are almost exactly the same. Nor do they vary much with the distance from the centre. Dividing the sound-area of the stronger British earthquakes into zones bounded by successive isoseismal lines, the percentages of persons who hear the sound before the shock are 67, 69, 70 and 63 in successives zones; while the percentages of those who hear the sound which follows the shock are respectively 38, 44, 41 and 36.

For the earthquakes of other countries, corresponding figures are not obtainable. But there can be no doubt that the sound usually precedes the shock. In countries where earthquakes are common and the phenomena are well known, the preliminary sound serves as a warning by which many persons are saved from a coming disaster.

The simplest explanation of the general precedence of the shock by the sound is that the sound-vibrations travel more rapidly than those which constitute the shock. But, if this were the case, the sound, with increasing distance from the origin, would be more generally heard before the shock and less frequently after it. The figures which have been given above offer no support whatever to this explanation. They seem rather to show that vibrations of all periods, whether perceptible as sound or as shock, travel with approximately the same speed. Any slight decline in the audibility of the sound before and after the shock is not more than might be expected owing to the increasing distance from the origin.

The relation which the extent of the sound-area

bears to that of the disturbed area varies greatly in different earthquakes. In the majority of strong and violent earthquakes, the sound-area occupies a region surrounding the centre, while the disturbed area extends beyond it in every direction. In Great Britain, the sound-area of strong earthquakes—those which disturb areas of more than 5000 square miles—is often considerable. The Hereford earthquake of 1896, for instance, was felt over an area of about 98,000 square miles, while the sound was heard over a district containing 70,000 square miles. On an average, the sound-area in strong earthquakes is 64 per cent., or roughly two-thirds, of the disturbed area. In other countries, the sound-area as a rule bears a smaller proportion to the disturbed area. In the Verny (Turkestan) earthquake of 1887, for instance, the disturbed area contained about 400,000 square miles and the sound-area about 132,000 square miles. The Neapolitan earthquake of 1857 was felt over all the Italian peninsula south of lat. 42°; the sound was heard only in a small area of about 3300 square miles immediately around the centre. Of another Italian earthquake, that of March 12, 1873, the disturbed area contained 227,000 square miles, and the sound-area 22,000 square miles.

While the sound-area in European earthquakes thus attains considerable dimensions, the case is very different in Japan. In one earthquake, that of

February 22, 1880, the shock was sensible to a distance of 120 miles, while sound-records come as a rule from places not more than 14 miles from the centre. Again, of the Japanese earthquakes which originated beneath the land during the years 1885–1892, 26·5 per cent. are recorded as accompanied by sound. For the submarine shocks of the same period, the percentage is only 0·8, and none of the earthquakes studied originated at a greater distance than 40 or 50 miles from the shore, while the centres of 93 per cent. of the total number were not more than 10 miles distant.

In earthquakes of a moderate degree of intensity, the disturbed areas of which range from a few hundred to a few thousand square miles, the disturbed area and sound-area practically coincide. In many slight earthquakes, with disturbed areas of one or two hundred square miles or less, the sound-area sometimes overlaps the disturbed area, as a rule only on one side, but sometimes in every direction. In the Helston earthquake of 1898, which will be referred to later (fig. 23), the sound-area overlapped the disturbed area on one side. The Comrie earthquake of 1898 was a very weak one; the sound-area contained but a few square miles, while the shock was only felt at one or two places within it.

The limiting case, in which the disturbed area vanishes, that is, in which the earth-sounds are heard without any accompanying shocks, is of considerable interest. As a general rule earth-sounds form part of the series of after-shocks of a great earthquake, or occur as intercalated members of a series of weak shocks. The mysterious noises known locally as Barisâl guns, mist-poeffeurs, marinas, etc., or under the general name of brontides, are possibly merely earth-sounds following in the wake, but after a prolonged interval of time, of some past and almost forgotten shock.

Most, if not all, great earthquakes include numerous earth-sounds among their attendant crowds of after-shocks, especially in and near the central regions. In Great Britain they are occasionally numerous, especially at the village of Comrie in Perthshire. After the strong shock of October 23, 1839, one observer at Comrie noted between this day and the end of 1841, 19 earthquakes, 25 tremors and 234 earth-sounds. In this district, and elsewhere, there is a complete continuity from earthquake to earth-sound. Every stage of the process is before us, from the strong earthquake in which the disturbed area extends in all directions beyond the sound-area, through the weak earthquake, in which the relations of the areas are reversed, down to the earth-sound, when the shock itself is imperceptible. We may therefore conclude that earthquakes and earth-sounds are manifestations, differing only in degree and in the mode in which we perceive them, of one and the same phenomenon.

One of the most significant phenomena of earthquake-sounds is the fact that the sound-area and the isoseismal lines of an earthquake are as a rule not concentric. In the slight earthquakes of this country, the excentricity of the sound-area is manifested by its overlapping the disturbed area in one direction. In the accompanying map of the Helston earthquake of 1898 (fig. 23), the continuous lines represent isoseismal lines of intensities 3 and 4. The outer dotted line indicates the boundary of the sound-area on the side on which it overlaps the disturbed area. The inner dotted line, which is concentric with the former, separates the places where the sound was very loud from those where it was distinctly fainter. Both curves, with respect to the isoseismal lines, are clearly displaced towards the northwest, while the boundary of the sound-area extends beyond that of the disturbed area at both ends of the longer axis. In this earthquake the originating fault, it is worthy of notice, is inclined to the southeast.

Of these phenomena, the general precedence of the sound and the excentricity of the sound-area with reference to the isoseismal lines are those which throw most light on the origin of the earthquake-sound. As the vibrations which form the sound and the shock travel with approximately the same velocity, it is evident, from the precedence of the sound, that the two sets of vibrations originate for the most part in different regions of the focus, and that the portion from which the sound-vibrations proceed lies outside

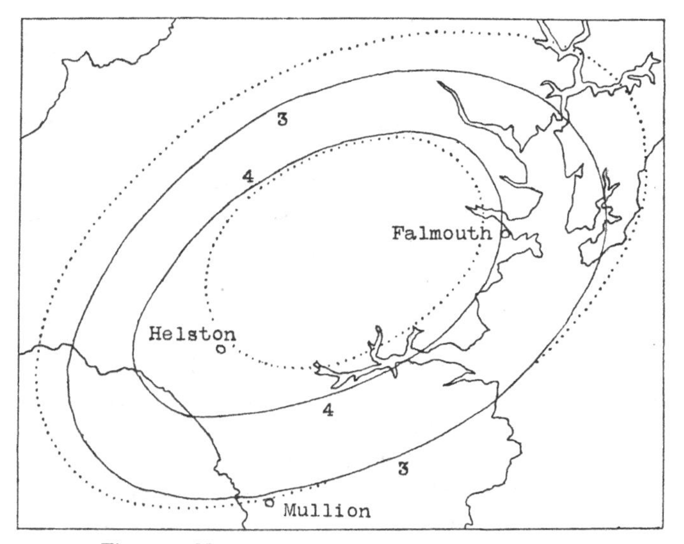

Fig. 23. Map of the Helston earthquake of 1898.

the other. The excentricity of the sound-area leads to the same conclusion. It implies that the origin of the sound-vibrations lies principally in the upper and lateral margins of the seismic focus, for the vibrations from the upper margin would be more audible than those, if any, which come from the lower margin.

The seismic focus is practically a surface inclined to the horizon. In its simplest form there is a central region of the fault-surface where the relative displacement of the two rock-masses is a maximum, and this is surrounded by a region in which the relative displacement is small and gradually dies away towards the edges. As the vibrations of greater range are also of long period, it is evident that from all parts of the focus there start together vibrations of various range and period, the large and slow vibrations coming mostly from the central region, and the small and rapid vibrations chiefly from its margins.

Now, between the sound-vibrations from the margins and the large vibrations from the central area which are only felt and not heard, there can be no discontinuity of period. Among the vibrations must therefore be included those which produce the deepest sound that can be heard by the human ear. It is evident also that the intensity of the sound must gradually increase until the shock is felt, after which it must gradually die away. Moreover, the sound from the nearer lateral margin of the focus will be heard by a larger number of persons than the sound from the farther margin, that is, the fore-sound will be heard by a larger percentage of observers than the after-sound. Lastly, the greater strength of the vibrations from near the central part of the focus will render audible vibrations of longer period than those which come from the margins, and thus the loud explosive crashes which are sometimes heard in the neigbourhood of the focus should accompany the strongest perceptible vibrations.

The magnitude of the sound-area depends chiefly on the dimensions of the seismic focus and therefore of its marginal regions. That of the disturbed area depends partly on the size of the focus, mostly perhaps on the initial intensity of the vibrations from its central portion. Thus, with very strong shocks, the sound-area may be a comparatively small district surrounding the centre. With very slight shocks, the marginal region may be so great compared with the central portion of the focus, that the sound-area may overlap the disturbed area. In the limiting case, the central portion of the focus would disappear, and a sound would be the only result sensible to human beings. Thus, the earth-sounds which are so prominent among the after-shocks of a great earthquake are merely the representatives of creeps along the fault-surface so small that they do not give rise to any vibrations that can be felt.

This explanation of the origin of earthquake-sounds throws light on several points in connexion with the growth of faults. Two of these may be mentioned in concluding this chapter.

The district represented in the accompanying map (fig. 24) is that part of Japan in which the great earthquake of 1891 and the majority of the numerous after-shocks were most prominently felt. During the eight years 1885–1892, the total number of earthquakes felt in the district was 3014, and of these 604, or 20 per cent., were accompanied by sound. The percentage, however, varies very widely in different parts of the area in question, and the continuous curves in the figure represent this variation. The meaning of any curve, say that marked 40, is as follows:—If any point on the curve be regarded as the centre of a small area, then 40 per cent. of the earthquakes the centres of which lie beneath this area were recorded as being accompanied by sound.

The dotted lines in fig. 24 bound the meizoseismal area of the earthquake, and within this area, again, the undulating line shows part of the path of the great fault-scarp. It will be noticed that the meizoseismal area is forked, the main branch proceeding towards the south, and that the principal group of curves follows this branch. The more northerly group of incomplete curves lies along the meizoseismal area, while the third group of curves lies roughly in the continuation of the easterly branch of that area.

In Japanese earthquakes the sound as a rule is heard only within a few miles from the origin. Other conditions, then, being the same, it follows that superficial earthquakes would have a greater chance of being audible than those which originate at a considerable depth, and hence that the curves of highest percentages in fig. 24 correspond to the earthquakes with the shallowest foci. The axes of

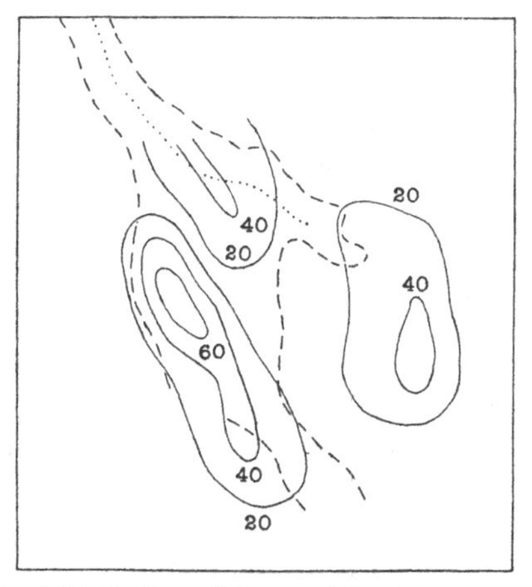

Fig. 24. Distribution of the audible after-shocks of the Mino-Owari earthquake of 1891.

the systems of curves thus mark out approximately the lines of growing faults, and indicate—what is not otherwise evident—that the displacement which gave rise to the fault-scarp is continued some miles farther to the southeast without producing any visible effect at the surface, and that displacements occurred, with a similar limitation, along another fault running along the main portion of the meizoseismal area.

In the third chapter it was shown that the after-shocks of the Inverness earthquake of 1901 proceeded from foci the depth of which gradually and continually decreased. A similar result follows from the varying percentage of earthquakes that were accompanied by sound in the districts mainly affected by the Mino-Owari earthquake of 1891. In the month of November (immediately after the great earthquake) that percentage was 18, and during the next five months it lay between 10 and 12; then it suddenly rose to 39 in May 1892, and during the following seven months it never fell below 32, while it maintained an average of 42. In certain smaller areas in the same district the same change is noticeable. In one, the percentage of audible earthquakes rose from 8 during the three months November 1891 to January 1892, to 39 during the next eleven months; in another, from 10 to 55.

Thus, both in moderately strong and in violent earthquakes, the strain produced by the displacement is increased in the portions of the fault immediately surrounding the focus, and especially in that above it. Little by little, by slip after slip, the strain is gradually relieved, until, even at the surface, it is no longer capable of causing the minute creeps which suffice to produce the phenomena of earthquake-sounds.

CHAPTER XII

DISTRIBUTION OF EARTHQUAKES

FROM a practical point of view the detailed study of the distribution of earthquakes is of no little importance; but, under the guidance of the theory described in the preceding pages, it acquires a fresh and loftier meaning. It shows us that the districts affected by earthquakes are precisely those in which the earth's crust is undergoing change. When we examine a seismic map of the world, we see at once the regions in which straits are growing wider and deeper, in which continents are extending and mountain-ranges are in process of coming to birth or sinking in decay. A country in which earthquakes have become slight and infrequent is one that has reached a stage of comparative old age. A region in which earthquakes are violent and constantly occurring is still in the condition of adolescence.

Of the two methods which have been devised for the construction of seismic maps, the older one of colouring the areas disturbed by all known earthquakes is much the less effective. It enables us to represent the distribution of seismic energy in broad lines, but it tends to obscure the essential details. Its chief advantage is that it depends on the records of thousands of earthquakes spread over a period of many centuries.

Of much greater significance and value are the maps on which the centres alone are marked. Except in detailed studies of special regions, minor shocks may be omitted with advantage. If we wish to ascertain the parts of the world which are undergoing change most rapidly, it is sufficient to limit ourselves to the great 'world-shaking' earthquakes, which are now recorded in hundreds of observatories scattered over the globe. Though many of these earthquakes occur beneath the sea or are unfelt in civilised countries, it is possible, from a comparison of the instrumental records, to determine approximately the positions of their origins. The great merit of the method is that no earthquakes of this class, wherever they occur, can escape registration. Its disadvantage is that our principal catalogue dates only from the year 1899. And the disadvantage is a serious one, for great earthquakes often re-visit a country formerly stricken only after the lapse of generations or even centuries.

The seismic map of the world given in fig. 25 is founded on the records collected by the Seismological Committee of the British Association in the years 1899–1909. During these eleven years, the number

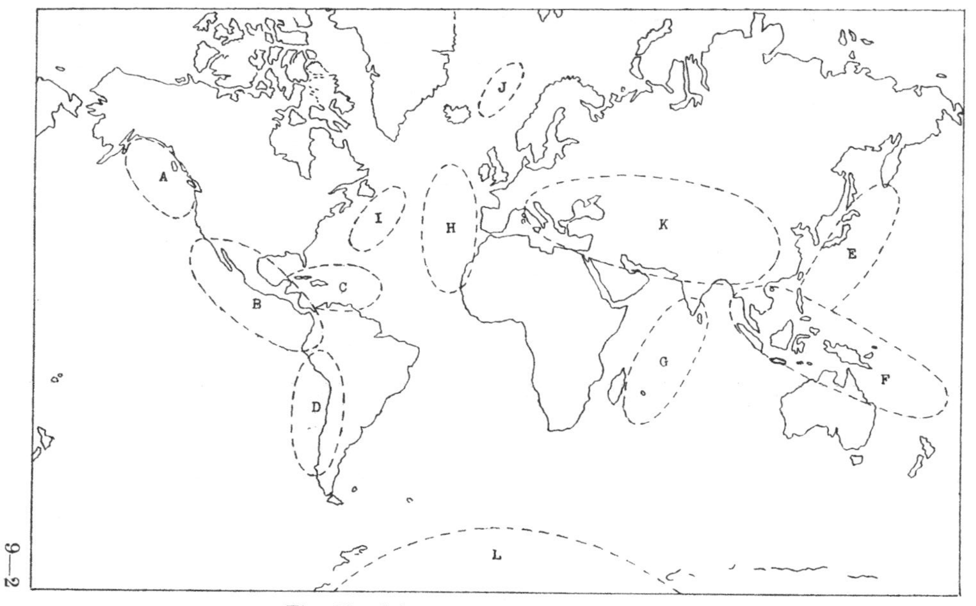

Fig. 25. Seismic map of the world. (Milne.)

of world-shaking earthquakes is 675, on an average 61 a year. The origins of all these earthquakes are known approximately. When plotted on a map they show a marked tendency to occur in groups. The long oval curves on the map represent roughly the boundaries of these areas of concentration, of which twelve have been defined. Four of them (*G–J*) are entirely oceanic, one (*K*) for the most part terrestrial, six (*A–F*) are partly oceanic and partly trench on the land, while one (*L*), of little consequence, is confined to the antarctic regions.

Of the four regions which are purely oceanic, one (marked *G*) in the Indian Ocean includes 26 of the 675 centres, another (*H*) to the west of Portugal contains 33. The other two regions (*I* and *J*) to the east of North America and the east of Greenland include 5 each. Thus, the four oceanic regions contain 10½ per cent. of all the earthquakes, the two regions *G* and *H* containing between them 9 per cent.

The region *C*, which includes the West Indies, a well-known unstable region, is responsible for 30 earthquakes, or 4½ per cent. of the total number.

The most interesting regions are the five which lie along the borders of the Pacific Ocean. Of these, *A* contains 40 centres, *B* 55, *D* 28, *E* 133 and *F* 175, altogether 431, or 64 per cent. of all the earthquakes considered. In other words, two out of every three world-shaking earthquakes originate under the steeply-sloping margins of the Pacific Ocean.

The remaining district (*K*), which is almost entirely terrestrial, is also of great importance, for 141 earthquakes, or 21 per cent., originated within it. The region includes such active seismic centres as Calabria, the Grecian archipelago, northern India and Turkestan, and such recent mountain-ranges as the Alps, the Caucasus and the Himalayas.

A map like that in fig. 25 is of necessity general in its character. But it indicates clearly the great law that governs the distribution of earthquakes, namely, that earthquakes are on the whole most frequent and violent in those regions of which the surface is steeply inclined to the horizon. Compare, for instance, in this respect the Atlantic and Pacific margins. The bed of the Atlantic slopes gently from either side, and the margins, except in the West Indies, are comparatively free from earthquakes. On the other hand, the borders of the Pacific are steeply inclined, the mountains of both North and South America line the western coast and their slopes are continued beneath the ocean, and it is precisely such regions which are among the most unstable of the globe. The great terrestrial region, extending from the Alps to beyond the Himalayas, is also marked in places by great surface-inclination.

When we examine a small district in detail, we find the same law of distribution manifested as clearly as it is on a large scale, and we also see some reason for its wide operation. Take, for instance, the chain of islands which compose the Japanese empire. They are arranged (fig. 26) in a festoon or curve convex towards the east and southeast. On the west side lies the Sea of Japan, into which the ocean-bed slopes gradually, the average gradient from the shore-line being 1 in 60 or less. On the east side, the ocean-bed shelves rapidly downwards into the well-known basin called the Tuscarora Deeps, the depth of which, one of the greatest recorded, is 4655 fathoms. The average gradient in this direction is about 1 in 30 and is in places as high as 1 in 20. Now, though earthquakes are not unknown on the west side of Japan, they are generally slight and comparatively infrequent. On the east side, they are numerous and many of them, like that of 1896, attain a high degree of intensity and are accompanied by destructive sea-waves.

During the eight years 1885–1892, the number of earthquakes recorded in Japan was 8331. Of these, 2488, or 30 per cent., had submarine origins. This is probably less than the actual number, for many of the shocks which originated at more than twenty miles from the shore would be imperceptible on land. The epicentres of the great majority of these submarine earthquakes have been determined approximately,

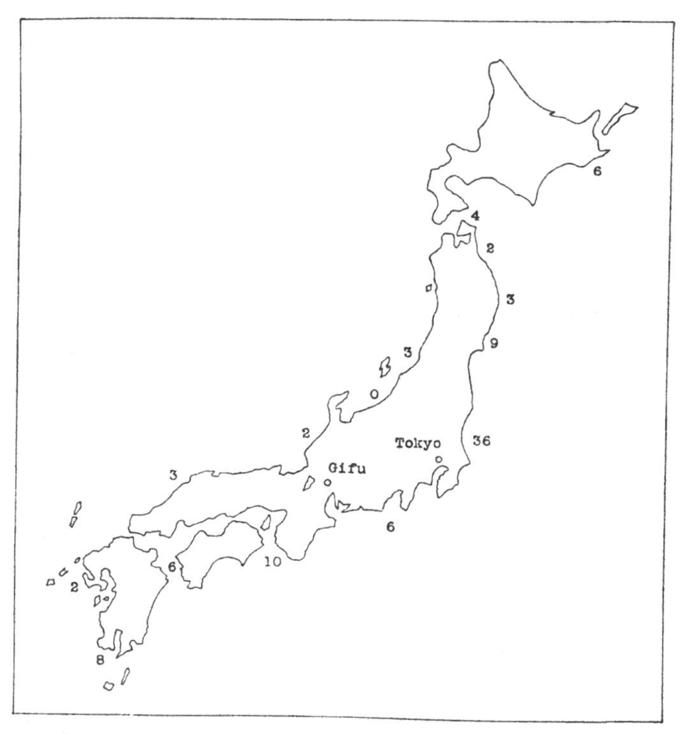

Fig. 26. Distribution of submarine earthquakes in Japan.

and they are found to occur in groups. No attempt is made in fig. 26 to define the boundaries of these districts, but figures are placed near their centres and indicate the percentage of submarine earthquakes which originated in the corresponding districts. It will be seen at once how numerous and important are the districts on the east coast compared with those on the west. Omitting the districts which belong neither to one coast nor the other, the total number of shocks that originated during the eight years on the east side was 1649, on the west side 172. That is to say, for every submarine earthquake which originated on the slightly shelving west coast of Japan, nine or ten originated on the east side which slopes rapidly into the Tuscarora Deeps.

The Japanese empire has been taken as the typical seismic district because in no other country have the earthquakes been studied and catalogued in such detail. Our knowledge of the steeply-sloping side of the festoon facing the Tuscarora Deeps is, however, limited to the form of its surface. We can only make any inference as to the structure from its analogy with similar seismic districts in a more advanced stage of growth, say, with a great mountain-system like the Himalayas. Here, the central ranges consist of ancient rocks which have been brought up to the surface and elevated into the loftiest peaks upon the globe. They are bounded, along the southern margin, by the sub-Himalayan chains composed for the most part of much more recent Tertiary beds and arranged in a regular curve, convex towards the south, on which side they slope steeply into the plains of India. The occurrence of great earthquakes, like that which ruined Kangra and Dharmsala in 1905, along the sub-Himalayan band, shows that the growth of the Himalayas is still far from being at an end, that the central masses are still, as of old, pressing forwards to the south, crushing and riding over the advanced guard of Tertiary mountains.

There can be little doubt that a similar advance is beginning to take place along the eastern coast of Japan, and that the numerous earthquakes which are felt in that district are indices of steps in the process which, in far distant ages yet to come, will culminate in a great mountain-range overlooking the Pacific Ocean.

To follow the process in greater detail would be to trench on another subject and a more difficult problem. In their effects on man and his works, earthquakes may seem at times to be all-powerful; but, as events in the evolution of the earth's crust, they sink into insignificance. The part they play in the formation of mountain-chains is but a small one; for they are merely the passing results of the movements by which those chains are raised.

But, if they occupy a lower place than was once thought among the operations of nature, earthquakes have at the same time gained in interest. To trace their distribution in time and space is to discover the laws which govern the growth of faults. When these laws are fully known—for we are as yet on the threshold of the subject—we shall have made no inconsiderable advance towards the solution of the great problem of the origin and growth of mountain-chains.

REFERENCES

CHAPTER I

BARATTA, M. Sopra le zone sismologicamente pericolose delle Calabrie. Voghera (N.D.).

CHAPTER III

DAVISON, C. The Inverness earthquake of September 18th. 1901, and its accessory shocks. Quart. Journ. Geol. Soc. vol. LVIII. 1902, pp. 377–397.

CHAPTER IV

1. DAVISON, C. The Derby earthquakes of March 24th and May 3rd, 1903. Quart. Journ. Geol. Soc. vol. LX. 1904, pp. 215–232.
2. DAVISON, C. The Derby earthquakes of July 3rd, 1904. Quart. Journ. Geol. Soc. vol. LXI. 1905, pp. 8–17.
3. DAVISON, C. Twin-earthquakes. Quart. Journ. Geol. Soc. vol. LXI. 1905, pp. 18–33.

CHAPTER V

The Californian earthquake of April 18, 1906 : Report of the State Earthquake Investigation Committee. Edited by A. C. Lawson. Two volumes and atlas. Washington, 1908–10.

CHAPTER VI

1. KOTO, B. The cause of the great earthquake in Central Japan, 1891. Journ. Coll. of Science, Imp. Univ. Japan, vol. V. 1893, pp. 295–353.
2. MILNE, J. A note on the great earthquake of October 28th, 1891. Brit. Assoc. Rep. 1892, pp. 114–128 ; Seis. Journ. of Japan, vol. I. 1893, pp. 127–151.

CHAPTER VII

OLDHAM, R. D. Report on the great [Assam] earthquake of 12th June 1897. Memoirs of the Geol. Survey of India, vol. XXIX. 1899, pp. 1–379.

CHAPTER VIII

MILNE, J. A catalogue of 8331 earthquakes recorded in Japan between 1885 and 1892. Seis. Journ. of Japan, vol. IV. 1895, pp. 1–367.

Many of the results in this and other chapters are based on the materials furnished by this great catalogue.

CHAPTER IX

1. DAVISON, C. On the distribution in space of the accessory shocks of the great Japanese earthquake of 1891. Quart. Journ. Geol. Soc. vol. LIII. 1897, pp. 1–15.
2. OLDHAM, R. D. List of after-shocks of the great [Assam] earthquake of 12th June 1897. Memoirs of the Geol. Survey of India, vol. XXX. 1900, pp. 1–102.
3. OMORI, F. On the after-shocks of earthquakes. Journ. Coll. of Science, Imp. Univ. Japan, vol. VII. 1894, pp. 111–200.

CHAPTER X

DAVISON, C. On the effect of the great Japanese earthquake of 1891 on the seismic activity of the adjoining districts. Geol. Mag. vol. IV. 1897, pp. 23–27.

CHAPTER XI

DAVISON, C. On earthquake-sounds. Phil. Mag. vol. XLIX. 1900, pp. 31–70.

CHAPTER XII

1. MILNE, J. Introduction to a catalogue of 8331 earthquakes recorded in Japan between 1885 and 1892. Seis. Journ. of Japan, vol. IV. 1895, pp. i–xxi.
2. MILNE, J. Reports of the Seismological Committee for 1909–1911. Brit. Assoc. Reports, 1909–1911.
3. MONTESSUS DE BALLORE, F. DE. Les Tremblements de Terre : Géographie Séismologique, Paris, 1906, Introduction, pp. 1–29.

INDEX